THE CARTOON GUIDE TO GENETICS

Also by Larry Gonick

The Cartoon History of the Universe
The Cartoon History of the United States
The Cartoon Guide to Physics (with Art Huffman)
The Cartoon Guide to the Computer

THE CARTOON GUIDE TO GENETICS

updated edition

GONICK

WHEELIS

LARRY GONICK
& MARK WHEELIS

HarperPerennial

A Division of HarperCollinsPublishers

FIRST HARPERPERENNIAL edition published 1991

The Library of Congress has cataloged the previous edition of this book as follows:

Gonick, Larry.
   The cartoon guide to genetics.

   "CO/416."
   Includes index.
   1. Genetics—Caricatures and cartoons.   I. Wheelis, Mark.   II.Title.
QH436.G66 1983   575.1′0207   82-48252
ISBN 0-06-460416-0 (pbk.)

ISBN 0-06-273099-1 (pbk.)

   93 94 95 RRD 10 9 8 7

IN ANCIENT TIMES...

OUR ANCESTORS HAD A FIRST-HAND KNOWLEDGE OF NATURE. IN THOSE DAYS, EVERYONE WAS A BIOLOGIST, AND THE WORLD WAS A CLASSROOM !!

IN THEIR EARLIEST GLIMMERINGS OF THOUGHT, IT'S SAID, PEOPLE MADE NO DISTINCTION BETWEEN *LIVING* AND *NON-LIVING* THINGS. EVERYTHING WAS SUPPOSED TO BE ALIVE, A FIT SUBJECT OF "BIOLOGICAL" RESEARCH.

THIS INCLUDED TREES...

THEY MOVE!

THEY MOVE!

...ANIMALS...

...AND THE VERY STONES THEMSELVES!

OW! THEY MOVE!

IN THE COURSE OF THEIR STUDIES, OUR ANCESTORS MUST HAVE NOTICED AN OBVIOUS FACT: SOME THINGS TENDED TO *REPRODUCE* THEMSELVES.

PEOPLE DID IT...

...MAMMOTHS DID IT...

...AND, TO THE PRIMITIVE MIND, IT MAY WELL HAVE SEEMED THAT EVEN ROCKS COULD "GIVE BIRTH" TO LITTLE PEBBLES!

MANY SCHOLARS BELIEVE THAT PRIMITIVE PEOPLE SAW NO CONNECTION BETWEEN REPRODUCTION AND SEX. THE NINE MONTHS BETWEEN CONCEPTION AND BIRTH WAS SUPPOSEDLY ENOUGH TO STYMIE THE SMARTEST STONE-AGER... AND WHAT DOES SEX HAVE TO DO WITH THE REPRODUCTION OF ROCKS ??!!

FOR WEEKS I'VE BEEN WATCHING, AND I DON'T THINK THEY DO IT...

WE MUST ADMIT, THIS THEORY LEAVES US SLIGHTLY SKEPTICAL. IT SEEMS POSSIBLE THAT MEN MIGHT HAVE MISSED THE CONNECTION, BUT COULD **WOMEN** HAVE OVERLOOKED WHAT WAS HAPPENING TO THEIR OWN BODIES ??!

EVER NOTICE ANYTHING FUNNY ABOUT BABIES AND SEX?

YES... YOU CAN'T HAVE ONE WITHOUT THE OTHER...

C'MON... DON'T BE SHY...

ENLIGHTENMENT CAME,
ACCORDING TO THIS THEORY,
WHEN PEOPLE FIRST
DOMESTICATED ANIMALS —
AND SAW THEIR REPRODUCTIVE
CYCLE CLOSE-UP AND OFTEN:
MATING IN ONE SEASON,
BIRTH IN ANOTHER.

YOU MEAN, *I*,
LIKE A SHEEP—
EEP... EEP...?

IT MUST HAVE COME
AS A GREAT SHOCK
TO DISCOVER THAT
MEN HAD SOMETHING
TO DO WITH MAKING
BABIES... IT'S SAID
TO HAVE CAUSED
BIG CHANGES IN
SOCIETY, SUCH AS
FATHER'S' DAY
PATERNITY SUITS,
MARRIAGE, AND THE
PATRIARCHY — BUT THIS
IS A BIOLOGY BOOK,
AND WE WON'T GO
INTO ALL THAT...

5

ALONG WITH THIS CAME THE NOTION THAT *LIKE BEGETS LIKE*— THE FIRST REALLY GENETIC IDEA..

AND SO BEGAN

PRACTICAL GENETICS,

OR "*SELECTIVE BREEDING.*" THE HERDERS BEGAN CONTROLLING THEIR ANIMALS' MATING, CHOOSING THE "BEST" SPECIMENS FOR REPRODUCTION, AND GETTING RID OF THE "WORST."

THIS IS UNNATURAL

RESULT?

A BREED OF PROUD, TOUGH, WILD ANIMALS WAS REDUCED TO SOMETHING DOCILE, WOOLY, AND *SHEEPISH!!*

COUSINS!! WHAT'S *HAPPENED* TO YOU?

=SIGH= A TRIUMPH OF PRACTICAL GENETICS...

AT THE SAME TIME, PEOPLE WERE DOMESTICATING *PLANTS*:

HEEL!! HEEL!!

EARLY FARMERS USED THE SAME METHODS AS THE ANIMAL HERDERS, WEEDING OUT UNDESIRABLE STRAINS AND PLANTING ONLY THE BEST SEEDS.

THIS HAPPENED ALMOST EVERYWHERE IN THE WORLD: SCRAWNY WEEDS AND GRASSES WERE GRADUALLY TURNED INTO RICH, PRODUCTIVE CROPS. RICE, WHEAT, BARLEY, AND DATES IN ASIA; CORN, SQUASH, TOMATOES, POTATOES, AND PEPPERS IN AMERICA; YAMS, PEANUTS, AND GOURDS IN AFRICA — ALL SPECIALLY IMPROVED BY HUMANS !!

HMM... EVERYWHERE BUT EUROPE, THAT IS...

PLANTS HAVE SEX, TOO... THEY'RE JUST LESS NOISY ABOUT IT THAN ANIMALS. EARLY ON, PEOPLE NOTICED THE IMPORTANCE OF *POLLINATION*: POLLEN DUST MUST LAND ON A FLOWER BEFORE IT CAN PRODUCE FERTILE SEEDS.

HoWEVEr—

THE EARLY FARMERS REALLY DIDN'T KNOW *WHY* POLLINATION WORKED—SO THEY ADDED SOME *MAGIC*, JUST TO BE ON THE SAFE SIDE...

THESE ARE ASSYRIAN PRIESTS, POLLINATING A DATE PALM, AROUND 800 B.C.

8

THIS COMBINATION OF SCIENCE AND MAGIC IS NICELY ILLUSTRATED BY A BIBLE STORY... *GENESIS, CHAPTER 30,* OR...

THE CASE OF JACOB'S FLOCK

...IN THIS STORY, THE PATRIARCH *JACOB* AGREES TO TEND THE FLOCK OF HIS FATHER-IN-LAW *LABAN.* AS PAYMENT, JACOB MAY TAKE ALL THE "SPECKLED AND SPOTTED" ANIMALS FOR HIMSELF, WHILE LABAN KEEPS THE PURE BLACK ONES. THE TWO GROUPS ARE NOT TO INTERBREED.

JACOB'S FLOCK   JACOB   LABAN'S FLOCK

THE BIBLE DESCRIBES JACOB'S *FERTILITY MAGIC* CAREFULLY: HE STRIPPED THE BARK FROM WILLOW RODS, AND "MADE THE WHITE APPEAR WHICH WAS IN THE RODS", THEN SET THEM NEAR THE WATERING HOLE.

THE IDEA BEHIND JACOB'S ACTION IS THAT *LIKE BEGETS LIKE*: BY SHOWING THE WHITE IN THE WILLOW RODS, HE WAS TRYING TO BRING OUT THE WHITE IN LABAN'S BLACK ANIMALS !! THIS IS CALLED *SYMPATHETIC MAGIC*...

I SUGGEST YOU READ THE ORIGINAL TO SEE WHY LABAN DESERVED THIS !

THE POINT, GENETICALLY SPEAKING, IS THIS: IN FACT, THE PURE BLACK ANIMALS *BORE SPECKLED OFFSPRING* — AND SO JACOB'S FLOCK INCREASED ! WHY ??

RECESSIVE GENES !

WE'LL COME BACK TO THIS LATER !

HERE WE SEE ACCURATE GENETIC OBSERVATION SIDE-BY-SIDE WITH A NEAR TOTAL LACK OF UNDERSTANDING.

LABAN CERTAINLY DIDN'T GET IT !!

YOU GET MY GOAT !!

10

SOME OTHER GENETIC ITEMS FROM ANCIENT HISTORY:

THE CHINESE DISCOVERED "WALTZING" MICE, A MUTATION WHICH CAUSES THE ANIMAL TO STAGGER AROUND IN CIRCLES.

THE HINDUS OBSERVED THAT CERTAIN DISEASES MAY "RUN IN THE FAMILY." MOREOVER, THEY CAME TO BELIEVE THAT CHILDREN INHERIT ALL THEIR PARENTS' CHARACTERISTICS. "A MAN OF BASE DESCENTS CAN NEVER ESCAPE HIS ORIGINS," SAY THE LAWS OF MANU...

XENOPHON, A GREEK, HAD THIS TO SAY ABOUT BREEDING HOUNDS:

11

SEVERAL OTHER GREEKS, THINKING MORE DEEPLY THAN XENOPHON, DEVELOPED THE FIRST REAL *THEORIES* OF *HEREDITY*— IN OTHER WORDS, THEY ADDRESSED THE QUESTION: "WHY DO CHILDREN RESEMBLE THEIR PARENTS?"

EXCEPT THE ONES WHO RESEMBLE THE MILKMAN?

ACTUALLY, ONE PHILOSOPHER, *SOCRATES*, WONDERED WHY THEY SOMETIMES *DON'T*... HE USED TO SAY THAT THE SONS OF GREAT STATESMEN WERE USUALLY LAZY AND GOOD FOR NOTHING... WE SHOULD ALWAYS BEAR THIS IN MIND, THAT NOT EVERY QUALITY IS INHERITED...

UNFORTUNATELY, BY SUCH UNFLINCHING HONESTY, SOCRATES PROVOKED THE ATHENIANS TO PUT HIM TO DEATH...

NEXT PHILOSOPHER!

THE MOST COHERENT
GREEK THEORY OF
HEREDITY WAS THAT
OF THE FAMOUS DOCTOR
HIPPOCRATES.

(HIPPOCRATIC)
OATH

HIPPOCRATES
RECOGNIZED THAT
THE MALE CONTRIBUTION
TO A CHILD'S
HEREDITY IS CARRIED
IN THE **SEMEN**.
BY ANALOGY, HE
ASSUMED THERE WAS
A SIMILAR FLUID IN
WOMEN.

JUST FOR
THE ✳❂#$
✳❂@¿
OF IT!

THESE FLUIDS, HE REASONED,
WERE MADE THROUGHOUT THE
BODY, AND THEN COLLECTED IN THE
REPRODUCTIVE ORGANS.

THE SEMEN
FROM THE FINGERS
HAD THE MATERIAL
TO MAKE MORE
FINGERS; THAT
FROM THE HAIR
MADE HAIR, ETC
ETC ETC...

AT CONCEPTION, A SORT OF
BATTLE OF THE FLUIDS TOOK PLACE,
AND WHETHER THE CHILD'S HANDS
WERE MORE LIKE MOM'S OR DAD'S
DEPENDED ON WHOSE FINGER-SEMEN
WON OUT !!

UNFORTUNATELY, THE GREEK WHOSE IDEAS MOST INFLUENCED LATER GENERATIONS WAS NOT HIPPOCRATES, BUT **ARISTOTLE.** WHEN IT CAME TO SCIENCE, ARISTOTLE NEVER LET HIS IGNORANCE STAND IN THE WAY OF HIS THEORIES!!

BIOLOGY? I CAN DO IT WITH MY EYES CLOSED!

ARISTOTLE — CALLED "THE PERIPATETIC" BECAUSE HE PACED WHILE HE LECTURED — BELIEVED THAT ALL INHERITANCE CAME FROM THE *FATHER*... THE MALE SEMEN, HE SAID, DETERMINED THE BABY'S FORM, WHILE THE MOTHER MERELY PROVIDED THE MATERIAL FROM WHICH THE BABY WAS MADE...

BUT, ARI — THEN WHERE DO *GIRLS* COME FROM?

14

YES, THERE WAS NO GETTING AROUND IT... THIS SEEMED TO IMPLY THAT ALL CHILDREN OUGHT TO BE BOYS... WHO KNOWS? MAYBE THIS REVEALED SOME SUBCONSCIOUS WISH OF ARISTOTLE'S... THE ANCIENT GREEKS DID VALUE BOYS MORE HIGHLY THAN GIRLS.

IN MY VERSION OF THE IDEAL STATE, ALL PHILOSOPHERS WOULD BE REQUIRED TO GET PREGNANT, AT LEAST ONCE...

BUT THE PHILOSOPHER COULD HARDLY IGNORE THE EXISTENCE OF FEMALE BABIES. HE PATCHED UP HIS THEORY BY DECLARING THEY WERE CAUSED BY "INTERFERENCE" FROM THE MOTHER'S BLOOD.

AND NOW, ON TO PHYSICS...

ONE PHILOSOPHER, **EMPEDOCLES**, THOUGHT THIS MIGHT RESULT FROM THE MOTHER'S GAZING LONGINGLY AT STATUES DURING PREGNANCY.

GREEK CIVILIZATION MAY HAVE PERISHED, BUT...

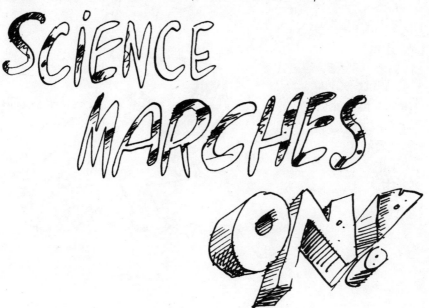

SCIENCE MARCHES ON!

THE GREEK MANTLE PASSED TO THE **ROMANS,** WHO HAD LITTLE TASTE FOR PHILOSOPHY,.. THEY PREFERRED THE TECHNOLOGY OF DEATH TO THE SCIENCE OF LIFE.

WHAT WAS YOUR CRIME?

PHILOSOPHY...

THE ONLY GENETIC IDEA THEY ADDED WAS THAT MARES COULD BE FERTILIZED BY THE WIND...

NEVER WORKED FOR ME... I MUST BE FACING THE WRONG WAY...

IN THE MIDDLE AGES,

SCIENCE FADED FURTHER... THEORIES OF HEREDITY GAVE WAY TO MERE LISTS OF "MONSTROUS" BIRTHS...

SOME OF THESE MAY WELL BE GENUINE — BUT WHAT ARE WE TO MAKE OF STORIES LIKE HALF A COW FALLING FROM HEAVEN IN A THUNDERCLAP?

THERE'S ALWAYS THE CHANCE IT WAS JUST A TALL TALE... OR SOMEONE'S IDEA OF A JOKE...

YES... REMINDS ME OF THE ONE ABOUT THE PRIORESS, THE ARCHDEACON, AND THE TWO-HEADED SWINE...

ONE MEDIEVAL IDEA ESPECIALLY IMPEDED UNDERSTANDING. IT WAS CALLED:

SPONTANEOUS GENERATION

IT'S COMMON SENSE!

ORIGINATING WITH THE GREEKS, THIS WAS THE BELIEF THAT *LIVING ORGANISMS* COULD ARISE ("SPONTANEOUSLY") FROM NON-LIVING MATTER.

MAMA!

MAGGOTS WERE SUPPOSED TO COME FROM DECAYING MEAT... HORSEHAIR TURNED INTO WORMS... AND FROGS, MICE, AND BUGS WERE NOTHING BUT SLIME COME TO LIFE!!

YOU CAN'T TELL ME OLD ARMOR DOESN'T BREED FLEAS!

IT'S NOT HARD TO IMAGINE WHY SPONTANEOUS GENERATION SEEMED PLAUSIBLE: IN A WORLD WHERE SLIME WAS COMMON, ONE SAW IT HAPPEN EVERY DAY!

YOU SEE HOW BELIEF IN SPONTANEOUS GENERATION CONFLICTS WITH "GENETIC" THINKING? IF A FROG COMES FROM SLIME, IT DOESN'T MAKE MUCH SENSE TO TALK ABOUT INHERITED QUALITIES, DOES IT??

NOT MUCH FAMILY RESEMBLANCE, IS THERE?

BUT — AS WE MENTIONED, SCIENCE MARCHES ON...

AND IN THE 17TH CENTURY, A SIMPLE EXPERIMENT SUCCESSFULLY CHALLENGED SPONTANEOUS GENERATION...

THE ELEGANT DEMONSTRATION WAS PERFORMED BY THE ITALIAN *FRANCESCO REDI*...

WHEN THE TIME IS RIGHT, THE MAN MUST BE REDI!

REDI PLACED PIECES OF FRESH MEAT IN JARS... SOME OF THE JARS HE CAPPED TIGHTLY WITH CHEESE-CLOTH, WHILE LEAVING THE REST OPEN TO THE FLIES...

AFTER SOME TIME HAD PASSED, REDI FOUND MAGGOTS ONLY IN THE OPEN JARS.

THE MAGGOTS GREW, STIFFENED INTO COCOONS, AND FINALLY EMERGED AS FULLY FORMED FLIES!

THUS, REDI HAD SHOWN THAT MAGGOTS COME FROM FLIES, AND FLIES COME FROM MAGGOTS. NOTHING VISIBLE HAD BEEN "SPONTANEOUSLY GENERATED" FROM THE ROTTING MEAT!!

BUT THE "SPONTANEOUS GENERATORS" WEREN'T DOWN YET...

SO-WE WERE WRONG ABOUT *FLIES*...SO WHAT?

PEOPLE STILL BELIEVED THAT FLEAS CAME FROM SAND, WEEVILS FROM GRAIN, EELS FROM THE DEW, ETC, ETC, ETC...

* *

FLEAS, EELS, AND WEEVILS, IN TURN, WERE DISPOSED OF BY *ANTON* VAN *LEEUWENHOEK* ("LAY-VEN-HOOK"), AN AMATEUR DUTCH SCIENTIST AND THE FIRST TO MAKE SYSTEMATIC USE OF THE *MICROSCOPE*.

WHERE DO HUMANS COME FROM?

HOSPITALS

USING HIS SIMPLE INSTRUMENT — JUST AN EXCELLENT EYEPIECE REALLY — LEEUWENHOEK FOLLOWED THE LIFE HISTORIES OF VARIOUS SMALL CREATURES. HIS TREATISE ON THE FLEA IS A CLASSIC!!

"THIS MINUTE AND DESPISED CREATURE," [HE WROTE] "IS ENDOWED WITH AS GREAT A PERFECTION IN ITS KIND AS ANY LARGE ANIMAL."

THANKS, TONY!

HE DISCOVERED THAT FLEAS, LIKE FISH, DOGS, AND HUMANS, WERE *SEXUAL BEINGS!*

MARK MY WORDS: FREE INQUIRY CAN ONLY LEAD TO FREE LOVE...

YES... LEEUWENHOEK HAS ALREADY CORRUPTED THE MORALS OF THE FLEA...

24

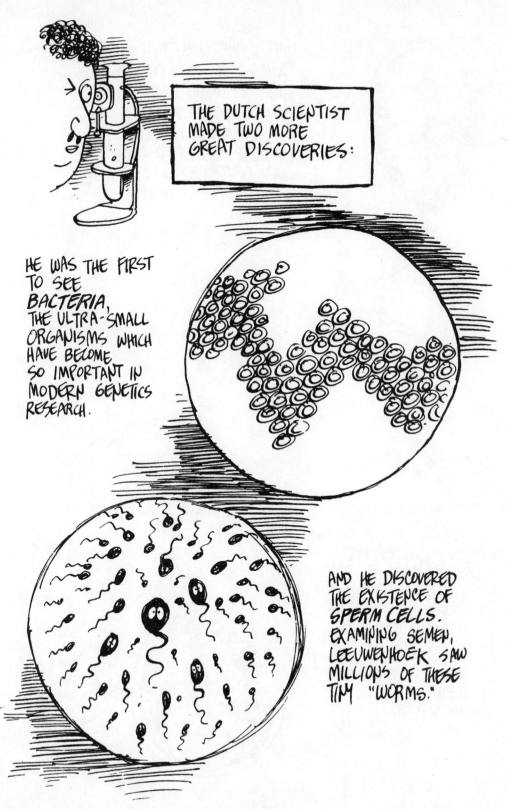

THE DUTCH SCIENTIST MADE TWO MORE GREAT DISCOVERIES:

HE WAS THE FIRST TO SEE *BACTERIA*, THE ULTRA-SMALL ORGANISMS WHICH HAVE BECOME SO IMPORTANT IN MODERN GENETICS RESEARCH.

AND HE DISCOVERED THE EXISTENCE OF *SPERM CELLS.* EXAMINING SEMEN, LEEUWENHOEK SAW MILLIONS OF THESE TINY "WORMS."

ONE MIGHT SAY THAT THIS DISCOVERY OPENED A WHOLE CAN OF WORMS... OR THAT IT SPAWNED WRONG IDEAS... FOR INSTANCE, LEEUWENHOEK HIMSELF BELIEVED EACH SPERM CELL CONTAINED A COMPLETE NEW ORGANISM IN MINIATURE.

THE OBVIOUS PROBLEM WAS: IF THIS "PRE-FORMED" ORGANISM WAS A BOY, IT MUST ALREADY HAVE TINY TESTICLES, WHICH WOULD CONTAIN MINIATURE SPERM, WHICH WOULD EACH HAVE EVEN TINIER PREFORMED ORGANISMS... AD INFINITUM ET ABSURDUM !!!

GET ME A TINY MICROSCOPE!!

EX OVO OMNIA

(AS LONG AS WE'RE TALKING LATIN!)

WHILE LEEUWENHOEK SPECULATED ABOUT SPERM, OTHER SCIENTISTS WERE LOOKING INTO THE FEMALE ROLE IN REPRODUCTION...

SIGNORA! LEMME SEE YOUR ORGANS! I'LL BE SCIENTIFIC...

"SHRIEK!" SIGNOR FALLOPIO! CONTROL YOURSELF!

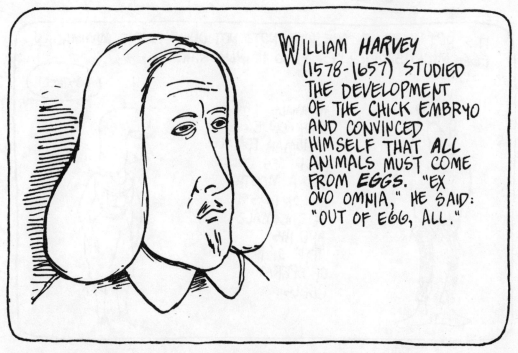

WILLIAM HARVEY (1578-1657) STUDIED THE DEVELOPMENT OF THE CHICK EMBRYO AND CONVINCED HIMSELF THAT ALL ANIMALS MUST COME FROM EGGS. "EX OVO OMNIA," HE SAID: "OUT OF EGG, ALL."

27

HARVEY BEGAN THE HUNT FOR THE *MAMMALIAN EGG.*

HE PERSUADED THE KING TO LET HIM LOOK FOR MAMMAL EGGS IN THE ROYAL DEER PARK... DOZENS OF DISSECTED DEER LATER, HARVEY HAD TO ADMIT FAILURE.

≡SIGH≡ GUESS I LAID AN EGG...

EX OVO OMELET!

FOR 200 YEARS THE HUNT WENT ON... AND STILL NO ONE COULD LOCATE THE ELUSIVE EGG.

IT'S NOT HARD TO SEE WHY NOT... NOT ONLY IS THE MAMMALIAN EGG MICROSCOPIC, IT'S ALSO FAIRLY RARE...

MAMMALS "LAY" VERY FEW EGGS: A HUMAN FEMALE PRODUCES ONLY ONE A MONTH, IN CONTRAST TO THE MALE AND HIS TENS OF MILLIONS OF SPERM CELLS.

WASTREL!

28

BUT THE SEARCH WENT ON... THERE WERE SOLID REASONS FOR BELIEVING MAMMALS HAD EGGS: WE HAVE OVARIES AND OVIDUCTS... IT WOULD BE PRETTY SILLY NOT TO HAVE EGGS, TOO...

YES, PEOPLE ARE JUST HIGHLY EVOLVED CHICKENS!

IN FACT, SCIENTISTS GREW SO SURE EGGS WERE THERE, THAT WHEN ONE WAS FINALLY SEEN—A DOG'S EGG, IN 1827— IT CAME AS MORE OF A RELIEF THAN A SURPRISE!!

HEY! DOGS GOT EGGS!

YOO-HOO!

OVER HERE!

I SAW IT!

HEY!

:SIGH: IT'S ABOUT TIME...

THE ONLY REMAINING RIDDLE WAS ANSWERED WHEN *OSCAR HERTWIG* OBSERVED THAT *FERTILIZATION* WAS THE UNION OF A SINGLE SPERM WITH A SINGLE EGG.

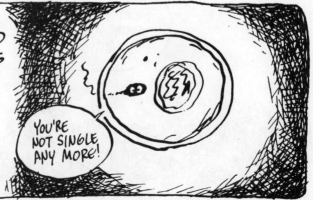

YOU'RE NOT SINGLE ANY MORE!

MEANWHILE,

SOME PROGRESS HAD BEEN MADE IN THE QUESTION OF *PLANT SEX.*

By 1700, THE SEXUAL NATURE OF PLANTS HAD BEEN LARGELY RESOLVED BY *CAMERARIUS* (1665-1721), WHOSE NAME EVEN SOUNDS LIKE A PLANT...

CAMERARIUS SHOWED THAT *FLOWERS* BORE SEX ORGANS QUITE LIKE THOSE OF ANIMALS.

AND THEY STICK THEM RIGHT IN THE AIR... SHAMEFUL!

30

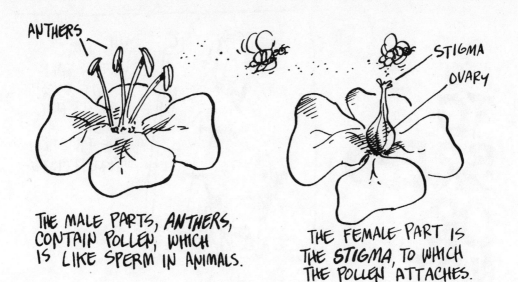

ANTHERS

STIGMA

OVARY

THE MALE PARTS, *ANTHERS*, CONTAIN POLLEN, WHICH IS LIKE SPERM IN ANIMALS.

THE FEMALE PART IS THE *STIGMA*, TO WHICH THE POLLEN ATTACHES.

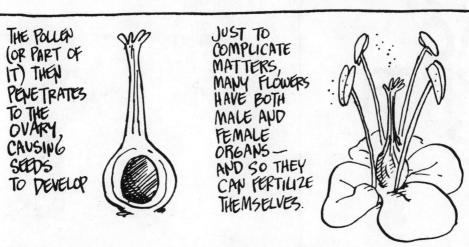

THE POLLEN (OR PART OF IT) THEN PENETRATES TO THE OVARY, CAUSING SEEDS TO DEVELOP

JUST TO COMPLICATE MATTERS, MANY FLOWERS HAVE BOTH MALE AND FEMALE ORGANS — AND SO THEY CAN FERTILIZE THEMSELVES.

SO BY THE EARLY 1800'S, BOTH PLANTS AND ANIMALS WERE KNOWN TO BE SEXUAL... THE MALE CONTRIBUTED POLLEN OR SPERM; THE FEMALE EGGS... AND SPONTANEOUS GENERATION WAS ON ITS LAST LEGS — ALMOST...

MY MOTHER SAID BABIES COME FROM CABBAGE LEAVES...

ARE YOU CALLING MY MOTHER A LIAR ??

TO BREED OR NOT TO BREED?

WITH ALL THIS TALK ABOUT SCIENTISTS, LET US NOT FORGET THE PRACTICAL GENETICISTS—

NAMELY, THE FARMERS AND STOCKBREEDERS WHO DID ALL THE DIRTY WORK OUT IN THE FIELDS.

SORRY!

FOR THEM, THE EARLY 19TH CENTURY WAS ALSO A TIME OF GREAT PROGRESS, WHEN PRACTICAL QUESTIONS OF FARMING WOULD LEAD, MORE OR LESS DIRECTLY, TO THE DISCOVERY OF THE GENE.

LET'S SEE WHAT THEY ALREADY KNEW FROM EXPERIENCE:

1. SOME *STABLE VARIETIES* NEARLY ALWAYS BREED TRUE, THEIR OFFSPRING HAVING THE SAME CHARACTERISTICS AS THEIR PARENTS. SOME COMMON EXAMPLES ARE MACKINTOSH APPLES, ARABIAN HORSES, LABRADOR RETRIEVERS, PEOPLE WITH BLUE EYES, ETC ETC ETC...

ON THE OTHER HAND, SOME BREEDING GROUPS SHOW GREAT VARIATION. JACOB'S FLOCK IS AN EXAMPLE OF VARIABLE COLOR. PEOPLE WITH BROWN EYES CAN HAVE BLUE-EYED CHILDREN.

2. IT IS SOMETIMES POSSIBLE TO MATE PARENTS FROM TWO DIFFERENT VARIETIES TO FORM **HYBRIDS.** FOR EXAMPLE, A MULE IS HALF HORSE AND HALF DONKEY. OF COURSE, NOT ALL HYBRIDS ARE POSSIBLE !!!

IMPOSSIBLE HYBRIDS:

PIG/TREE   HUMAN/STRAWBERRY

HYBRIDS ARE DIFFICULT TO PREDICT... THEY MAY SEEM VIRTUALLY IDENTICAL TO ONE PARENT, OR THEY MAY COMBINE FEATURES OF BOTH — AND WHEN HYBRIDS BREED WITH HYBRIDS, THE RESULT IS VARIATION IN THE EXTREME!!

HARD TO BELIEVE YOU'RE MY BROTHER!

3. ALL VARIETIES, EVEN STABLE ONES, OCCASIONALLY PRODUCE "*SPORTS*" — OFFSPRING DIFFERENT FROM EITHER PARENT. THESE ARE OFTEN GROSSLY DEFECTIVE "MONSTROSITIES"...

OUR CHILD IS A MESS!

A DACHSHEEP!

BUT SOMETIMES THE SPORT DIFFERS ONLY SLIGHTLY, LIKE THE STUBBY-LEGGED SHEEP WHICH APPEARED AROUND 1800.

BY CROSSING THESE SPORTS BACK WITH NORMAL TYPES, 19TH-CENTURY FARMERS HAD MANAGED TO CREATE SEVERAL NEW STABLE VARIETIES. THERE WERE NEW TYPES OF WHEAT, PEAS, AND STRAWBERRIES, HORNLESS CATTLE, AND STUBBY-LEGGED SHEEP.

BUT IT WAS STILL A MATTER OF TRIAL AND ERROR... IT DIDN'T ALWAYS WORK... AND SO PEOPLE BEGAN TO WONDER IF THERE MIGHTN'T BE A SCIENTIFIC WAY OF SELECTING ADVANTAGEOUS TRAITS TO CREATE NEW VARIETIES.

IF WE COULD BREED A SIX-LEGGED HORSE, WE'D CLEAN UP IN GLUE!

AND A THREE-LEGGED HUMAN COULD PUT HIS FOOT IN HIS MOUTH AND STILL WALK!

HOWEVER,

DESPITE A GOOD DEAL OF WORK, NO TRULY GENERAL LAWS OF INHERITANCE WERE DISCOVERED.

SOME INVESTIGATORS CONFUSED THEMSELVES BY CROSSING BREEDS THAT DIFFERED IN TOO MANY CHARACTERISTICS...

OTHERS FAILED TO KEEP A CAREFUL COUNT OF THE NUMBER OF VARIETIES PRODUCED FROM EACH CROSS.

*@# FLEAS ARE HARD TO KEEP TRACK OF!

INDEED, THE PROBLEM SEEMED HOPELESS... GRADUALLY, SCIENTISTS GAVE UP TRYING AND TURNED TO EASIER PROBLEMS... AND THAT IS WHY, WHEN THE LAWS OF INHERITANCE WERE FINALLY FIGURED OUT, THE DISCOVERY WAS IGNORED FOR THIRTY YEARS...

MONK FINDS GENE; WORLD YAWNS!

Ho Hum!

FIFTY YEARS OF RESEARCH HAD FAILED TO FIND ANY PRECISE LAW OF INHERITANCE. OBVIOUSLY, DISCOVERING THE RIGHT FORMULA, IF POSSIBLE, WAS A JOB REQUIRING SUPERHUMAN PATIENCE, UNLIMITED TIME, AND, AS IT HAPPENED, A MIRACLE OF LUCK.

NO WONDER IT HAPPENED IN A MONASTERY...

GREGOR MENDEL

(1822 - 1884) WAS AN AUGUSTINIAN MONK FROM BRÜNN, AUSTRIA. IN HIS SPARE TIME, MENDEL BRED PEA PLANTS IN THE MONASTERY GARDENS.

BUT MENDEL WAS NOT JUST AN AMATEUR GARDENER, BUT A **SCIENTIST** WHO STUDIED HIS PEA PLANTS MOST CAREFULLY — HE CALLED THEM HIS "CHILDREN."

WHAT KIND OF DAD DOES EXPERIMENTS ON HIS KIDS?

CHOOSING PEAS WAS THE MIRACLE OF LUCK: THEY ARE PERFECTLY SUITED TO GENETIC RESEARCH, WITH A NUMBER OF STABLE VARIETIES WHICH MAY FORM HYBRIDS:

THERE WAS A TALL VARIETY AND A SHORT ONE...

ONE TYPE MADE SMOOTH, ROUND PEAS, WHILE ANOTHER'S WERE LUMPY AND WRINKLED...

SOME PODS WERE PLUMP, WHILE OTHERS WERE PINCHED...

THERE WERE GREEN PEAS AND YELLOW; GREY SEED-COATS AND WHITE; WHITE FLOWERS AND PURPLE. THERE WERE DIFFERENCES IN THE COLOR OF THE UNRIPE PODS, THE COLOR OF SEED ALBUMIN, AND THE POSITION OF THE FLOWERS.

EVERY PEA FLOWER HAS BOTH MALE AND FEMALE ORGANS, SO THEY ORDINARILY FERTILIZE THEMSELVES.

UNLESS WE PRACTICE ≥AHEM≤ FAMILY PLANNING!

HOW MENDEL MADE HYBRIDS:

FIRST HE SNIPPED OFF THE ANTHERS WHILE STILL IMMATURE TO PREVENT "SELFING."

THEN HE DUSTED THE STIGMA WITH POLLEN TAKEN FROM THE DESIRED "FATHER."

FINALLY, HE TIED BAGS OVER THE FLOWERS TO KEEP OUT ANY STRAY POLLEN.

IN THIS WAY MENDEL WAS ABLE TO CONTROL THE PARENTAGE OF EACH GENERATION.

PSST! I THINK THE MONK IS PLAYING GOD!!

MENDEL'S FIRST MAJOR RESULT WAS THE DISCOVERY OF *DOMINANCE*. WHAT HAPPENED WHEN A TALL PLANT WAS CROSSED WITH A SHORT? ONE MIGHT EXPECT MEDIUM-SIZED PLANTS, **BUT**

IN FACT, ALL THE HYBRIDS WERE *TALL!!*

MENDEL EXPRESSED THIS BY SAYING THAT TALLNESS WAS *DOMINANT* OVER SHORTNESS (IN PEAS!). THE TRAIT OF SHORTNESS IS THEN CALLED *RECESSIVE*. IN EVERY CASE, ONE TRAIT WAS FOUND TO BE DOMINANT.

ROUND SEEDS ARE DOMINANT OVER WRINKLED; PLUMP PODS OVER PINCHED; GREY SEED-COATS OVER WHITE SEED-COATS, ETC ETC ETC....

IT DIDN'T MATTER WHICH PARENT CONTRIBUTED THE POLLEN AND WHICH THE EGG. A TALL-SHORT HYBRID WAS ALWAYS TALL.

THE FUN BEGINS WHEN YOU START BREEDING THE HYBRIDS—

40

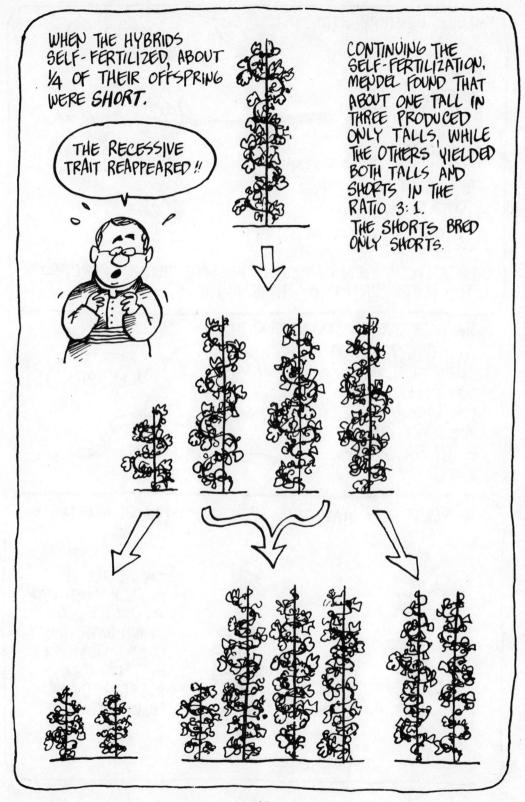

WHEN THE HYBRIDS SELF-FERTILIZED, ABOUT ¼ OF THEIR OFFSPRING WERE *SHORT*.

THE RECESSIVE TRAIT REAPPEARED !!

CONTINUING THE SELF-FERTILIZATION, MENDEL FOUND THAT ABOUT ONE TALL IN THREE PRODUCED ONLY TALLS, WHILE THE OTHERS YIELDED BOTH TALLS AND SHORTS IN THE RATIO 3:1. THE SHORTS BRED ONLY SHORTS.

MENDEL'S INTERPRETATION:

IT'S MATHEMATICAL!

THERE IS SOMETHING IN POLLEN AND EGG WHICH DETERMINES THE HEIGHT OF PEA PLANTS. THIS "SOMETHING" WE CALL A

GENE.

EACH POLLEN GRAIN AND EGG HAS ONE HEIGHT GENE, SO THE PLANT FORMED BY THEIR UNION HAS **TWO**.

THE GENE MAY BE ONE OF TWO DISTINCT TYPES, OR

ALLELES.

ONE ALLELE, A, IS FOR TALLNESS; THE OTHER ONE, a, IS FOR SHORTNESS.

GENES MAKE SHORTS?

CUT-OFFS...

A PLANT MAY HAVE THE SAME OR DIFFERENT ALLELES.

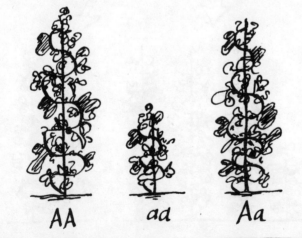

AA   ad   Aa

THE ALLELE A IS DOMINANT OVER a. THAT IS, THE PLANT WITH THE COMBINATION Aa IS TALL. THE ALLELES DO NOT "BLEND."

WHAT HAPPENS WHEN **AA** BREEDS WITH **AA**? POLLEN AND EGG EACH GET ONE COPY OF THE GENE... IN THIS CASE, THE ALLELES ARE THE SAME — A — SO THE OFFSPRING WILL AGAIN BE AA, OR TALL. LIKEWISE, aa CAN YIELD ONLY aa. THESE ARE THE STABLE SHORT & TALL VARIETIES.

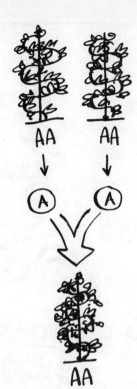
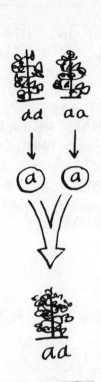

MENDEL'S FIRST HYBRID WAS A CROSS BETWEEN **AA** AND **aa**: THE POLLEN (OR EGG) FROM AA CONTAINS ONLY A, WHILE THE EGG (OR POLLEN) FROM aa CONTAINS ONLY a.

RESULT:

Aa, WHICH IS TALL.

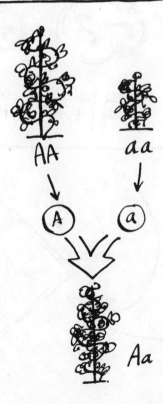

43

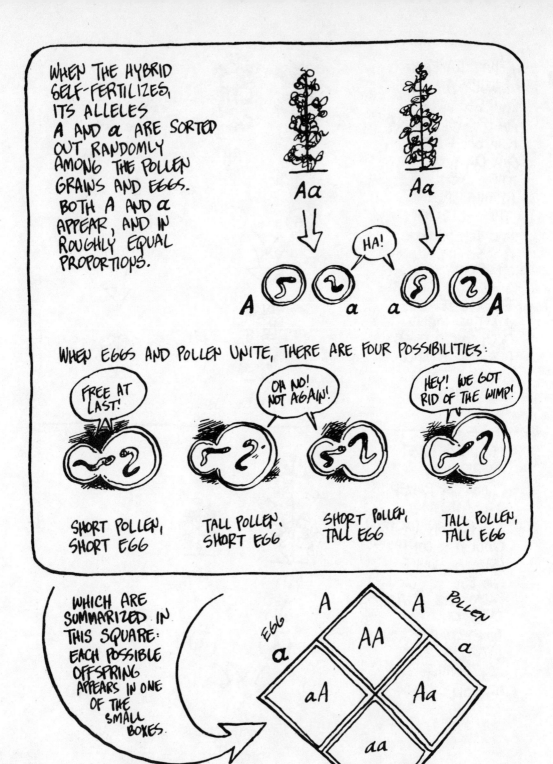

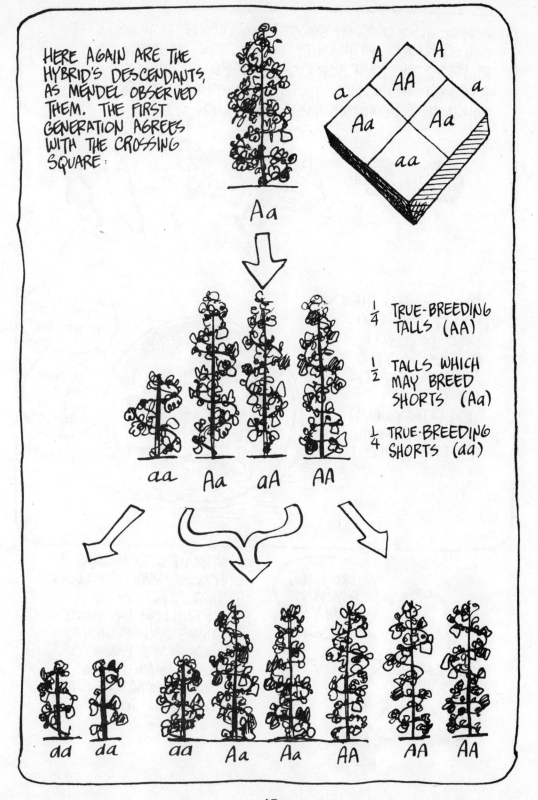

HERE AGAIN ARE THE HYBRID'S DESCENDANTS, AS MENDEL OBSERVED THEM. THE FIRST GENERATION AGREES WITH THE CROSSING SQUARE:

Aa

aa Aa aA AA

$\frac{1}{4}$ TRUE-BREEDING TALLS (AA)

$\frac{1}{2}$ TALLS WHICH MAY BREED SHORTS (Aa)

$\frac{1}{4}$ TRUE-BREEDING SHORTS (aa)

dd da aa Aa Aa AA AA AA

45

MENDEL ALSO CROSSED SMOOTH-PEA PLANTS WITH WRINKLED, PURPLE FLOWERS WITH WHITE, ETC ETC ETC. IN EVERY CASE, HE FOUND THE CHARACTERISTIC TO BE CONTROLLED BY A SINGLE GENE WITH TWO DIFFERENT ALLELES, ONE OF WHICH WAS DOMINANT OVER THE OTHER.

SO IT SEEMED THAT POLLEN AND EGG WERE BOTH FULL OF THESE LITTLE "SOMETHINGS," ONE FOR EVERY HEREDITARY TRAIT OF THE ORGANISM. PRETTY CROWDED!

HOW CAN I DO MY JOB IN THIS MOB?

YOU DON'T HAVE TO: YOU'RE RECESSIVE!

LORD KNOWS THEY MUST BE TINY!!

WITHOUT EVER SEEING A GENE, MENDEL CONCLUDED THAT HEREDITY IS CONTROLLED BY THESE "ATOMS OF INHERITANCE," WHICH NEVER BREAK OR BLEND, MAINTAINING THEIR CHARACTER FROM GENERATION TO GENERATION.

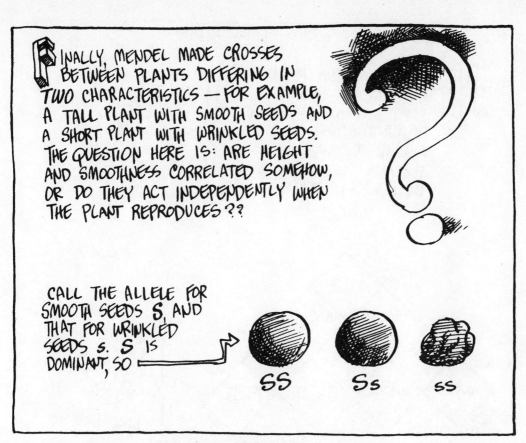

INALLY, MENDEL MADE CROSSES BETWEEN PLANTS DIFFERING IN TWO CHARACTERISTICS — FOR EXAMPLE, A TALL PLANT WITH SMOOTH SEEDS AND A SHORT PLANT WITH WRINKLED SEEDS. THE QUESTION HERE IS: ARE HEIGHT AND SMOOTHNESS CORRELATED SOMEHOW, OR DO THEY ACT INDEPENDENTLY WHEN THE PLANT REPRODUCES??

CALL THE ALLELE FOR SMOOTH SEEDS S, AND THAT FOR WRINKLED SEEDS s. S IS DOMINANT, SO

SS   Ss   ss

THE CROSS IS BETWEEN AASS AND aass.

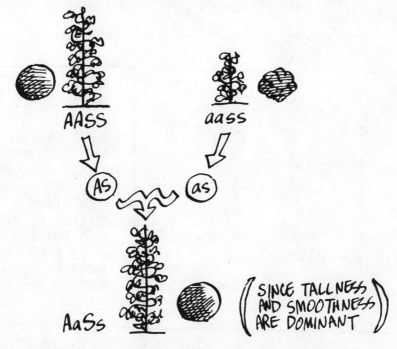

AASS   aass

AS   as

AaSs   (SINCE TALLNESS AND SMOOTHNESS ARE DOMINANT)

47

NOW FOR THE SELF-POLLINATION OF THE HYBRID:

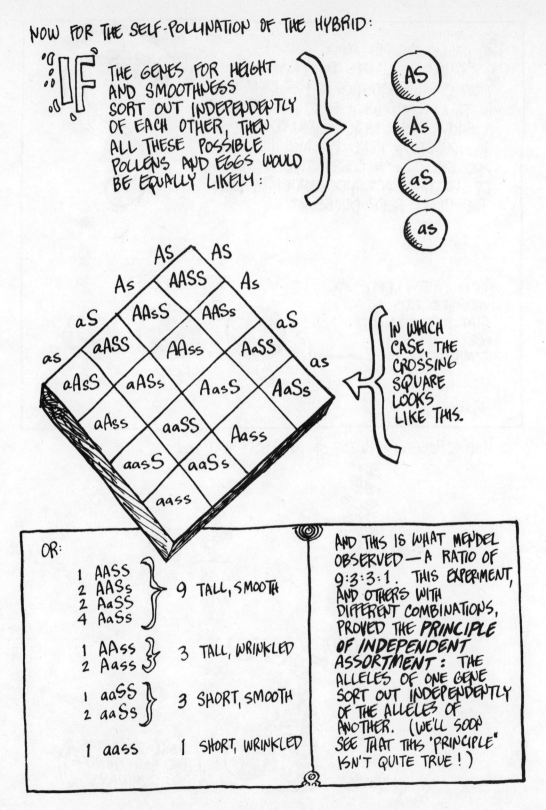

IF THE GENES FOR HEIGHT AND SMOOTHNESS SORT OUT INDEPENDENTLY OF EACH OTHER, THEN ALL THESE POSSIBLE POLLENS AND EGGS WOULD BE EQUALLY LIKELY:

As

As

aS

as

IN WHICH CASE, THE CROSSING SQUARE LOOKS LIKE THIS.

	AS	As	aS	as
AS	AASS	AASs	AaSS	AaSs
As	AASs	AAss	AaSs	Aass
aS	AaSS	AaSs	aaSS	aaSs
as	AaSs	Aass	aaSs	aass

OR:

1 AASS
2 AASs
2 AaSS
4 AaSs
} 9 TALL, SMOOTH

1 AAss
2 Aass
} 3 TALL, WRINKLED

1 aaSS
2 aaSs
} 3 SHORT, SMOOTH

1 aass   1 SHORT, WRINKLED

AND THIS IS WHAT MENDEL OBSERVED — A RATIO OF 9:3:3:1. THIS EXPERIMENT, AND OTHERS WITH DIFFERENT COMBINATIONS, PROVED THE **PRINCIPLE OF INDEPENDENT ASSORTMENT**: THE ALLELES OF ONE GENE SORT OUT INDEPENDENTLY OF THE ALLELES OF ANOTHER. (WE'LL SOON SEE THAT THIS "PRINCIPLE" ISN'T QUITE TRUE!)

NOW THAT WE'VE SEEN HOW GENES WORK, HERE'S A BIT OF GENETICS JARGON, IN CASE YOU SHOULD EVER WANT TO EAVESDROP ON A MODERN GENETICIST...

THIS GEN-TEK DEAL MEANS *ELEPHANT BUCKS,* BABY... WE'RE TALKING RECOMBINANT BANK ACCOUNTS, PROFESSOR...

WELL... NOT *THAT* KIND OF JARGON...

GENETICISTS DISTINGUISH BETWEEN AN ORGANISM'S *PHENOTYPE* — WHAT IT LOOKS LIKE — AND ITS *GENOTYPE* — WHAT ALLELES IT HAS.

AA

Aa

SAME PHENOTYPE, DIFFERENT GENOTYPE

AN ORGANISM IS *HOMOZYGOUS* WITH RESPECT TO A GIVEN GENE IF ITS TWO ALLELES ARE THE *SAME,* AND *HETEROZYGOUS* IF THEY'RE DIFFERENT.

SS
HOMOZYGOUS

Ss
HETEROZYGOUS

SO NOW YOU KNOW WHAT A GENETICIST MEANS BY "PHENOTYPICALLY SMOOTH, GENOTYPICALLY HETEROZYGOUS."

YES... NOW TELL ME ABOUT RECOMBINANT BANK ACCOUNTS...

INCIDENTALLY — WE'RE NOW IN A POSITION TO UNDERSTAND JACOB'S SPECKLED FLOCK:

SPELL IT OUT FOR ME!

THE ALLELE FOR A BLACK COAT, CALL IT B, WAS DOMINANT. THERE WAS ALSO A RECESSIVE ALLELE, w, FOR WHITE SPECKLES. MANY OF LABAN'S PHENOTYPICALLY BLACK ANIMALS SECRETLY HARBORED THIS w, SO THEIR KIDS WERE SOMETIMES SPECKLED.*

Bw × Bw

BB   Bw   Bw   ww

IN OTHER WORDS —

THOSE GOATS WERE HETEROZY-GOATS!

* ACTUALLY, THE GENETICS OF COAT COLOR ARE MORE COMPLEX, BUT THE PRINCIPLE IS THE SAME: RECESSIVE ALLELES.

QUESTION:

IF YOU SEE A DOMINANT PHENOTYPE, HOW CAN YOU TELL IF IT'S A HETEROZYGOTE?

IS IT POLITE TO ASK?

FOR INSTANCE, IN HUMANS BROWN EYES ARE DOMINANT OVER BLUE. CALL THE GENES B AND b, RESPECTIVELY.

HOW CAN WE TELL IF THIS BROWN-EYED PERSON IS BB OR Bb?

ONE WAY IS TO CROSS HIM WITH A RECESSIVE HOMOZYGOTE— I.E., A BLUE-EYED PERSON, bb.

SORRY... I HAVE TO BACK OUT OF THIS EXPERIMENT... MONK'S VOWS, YOU KNOW...

O.K... WE'LL USE SOMEBODY ELSE...

BROWN

BLUE

IF ANY OF THE LITTLE HYBRIDS HAS BLUE EYES, THE BROWN-EYED PARENT MUST HAVE BEEN A *HETEROZYGOTE*, Bb. IF HE HAD BEEN BB, ALL THE CHILDREN WOULD HAVE BEEN Bb, WITH BROWN EYES.

FOR EXAMPLE, MY FIRST WIFE HAS BROWN EYES, AND I HAVE BLUE EYES. ONE OF OUR SONS HAS BLUE EYES; ONE HAS BROWN EYES. THEREFORE, MY FIRST WIFE MUST BE *HETEROZYGOUS.* (THE BLUE-EYED BOY MUST HAVE ONE ALLELE FROM HER.) TRY DOING THE SQUARE!

MY SECOND WIFE HAS BLUE EYES LIKE ME. IF OUR CHILD HAD *BROWN* EYES, WHAT WOULD WE MAKE OF THAT? BETTER ASK THE MILKMAN!!

SOME EXAMPLES OF DOMINANT AND RECESSIVE GENES IN HUMANS:

✭ BROWN EYES ARE DOMINANT OVER BLUE EYES.

✭ COLOR VISION IS DOMINANT OVER COLOR BLINDNESS.

✭ HAIRY HEADS ARE DOMINANT OVER BALD ONES.

✭ THE ABILITY TO CURL THE TONGUE IS DOMINANT OVER THE INABILITY TO CURL THE TONGUE.

✭ EXTRA FINGERS ARE DOMINANT OVER FIVE FINGERS (ODD BUT TRUE!).

A DOUBLE DOSE OF RECESSIVES ALSO CAUSE SUCH RARE DISEASES AS HEMO-PHILIA, SICKLE-CELL ANEMIA, TAY-SACHS SYNDROME, THALASSEMIA, DWARFISM...

TO SUM UP...

1. 2. 3. 4. 5. 6.

MY PRINCIPAL RESULTS:

1 HEREDITARY TRAITS ARE GOVERNED BY GENES WHICH RETAIN THEIR IDENTITY IN HYBRIDS. GENES ARE NEVER BLENDED TOGETHER.

NO COMPROMISE WITH RECESSIVES!

2 ONE FORM ("ALLELE") OF A GENE MAY BE **DOMINANT** OVER ANOTHER. BUT RECESSIVE GENES WILL POP UP LATER!!

THE SECRET OF MY SPECKLED GOATS!

3. EACH ADULT ORGANISM HAS TWO COPIES OF EACH GENE — ONE FROM EACH PARENT. WHEN POLLEN OR SPERM AND EGGS ARE PRODUCED, THEY EACH GET ONE COPY.

4 DIFFERENT ALLELES ARE SORTED OUT TO SPERM AND EGG RANDOMLY AND INDEPENDENTLY. ALL COMBINATIONS OF ALLELES ARE EQUALLY LIKELY:

AABBCCDDEEFFGGHH
AaBBCCDDEEFFGGHH
aABBCCDDEEFFGGHH
aaBBCCDDEE FFGGHH
AAbBCCDDEE GGHH
AABbCCDDEE GGHH
AaBbCC GGHH
aABb GGHH

ETC!

➤➤➤ WE'LL SEE SHORTLY THAT NOT ALL THESE POINTS ARE EXACTLY CORRECT... DOMINANCE IS SOMETIMES ONLY PARTIAL... THERE ARE ORGANISMS WITH ONLY A SINGLE SET OF GENES... AND SOME WITH FOUR SETS... AND DEVIATIONS FROM INDEPENDENT ASSORTMENT TURN OUT TO BE VERY IMPORTANT...

MENDEL PRESENTED HIS THEORY IN 1865 TO THE BRÜNN NATURAL SCIENCE SOCIETY... IT PUT THEM TO SLEEP.

I CAN'T TAKE THE MATH...

UNFORTUNATELY, NOBODY CARED ABOUT THE PROBLEM ANY MORE... IT HAD GONE OUT OF FASHION... AND, BESIDES, SINCE 1859, BIOLOGISTS HAD BEEN DISTRACTED BY THE NEW THEORY OF *EVOLUTION*, AND COULDN'T BE BOTHERED WITH MENDEL'S EQUATIONS.

DARWIN

BY THE TIME MENDEL DIED, THE SCIENTIFIC COMMUNITY HAD TOTALLY FORGOTTEN HIS WORK. "MY TIME WILL COME," HE SAID, NOT LONG BEFORE HIS DEATH IN 1884...

GARDENING IS OUT OF FASHION THESE DAYS...

SOME FIELDS NEVER GO OUT OF FASHION...

NOW YOU SEE THEM...

WHILE MENDEL'S WORK LAY NEGLECTED, OTHERS WERE FINDING WONDERFUL THINGS IN THE MICRO-WORLD.

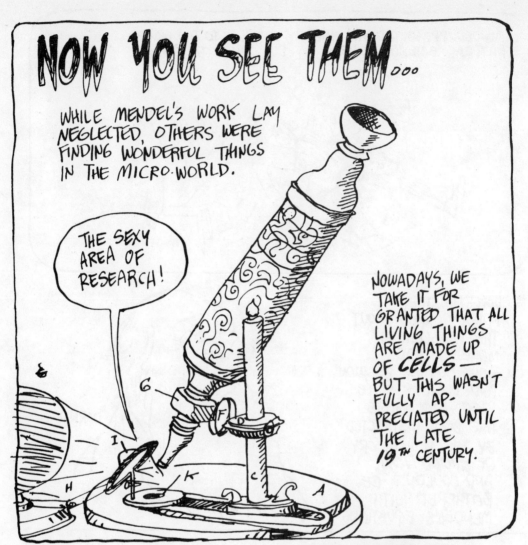

THE SEXY AREA OF RESEARCH!

NOWADAYS, WE TAKE IT FOR GRANTED THAT ALL LIVING THINGS ARE MADE UP OF **CELLS** — BUT THIS WASN'T FULLY AP- PRECIATED UNTIL THE LATE 19TH CENTURY.

AS FAR BACK AS THE 1600'S, ROBERT HOOKE (1635-1703) HAD NOTICED THE CELLULAR STRUCTURE OF CORK. BUT IT WASN'T UNTIL THE 1800'S THAT SCIENTISTS, ARMED WITH BETTER MICROSCOPES, REALIZED THAT ALL OF US ARE DIVIDED INTO LITTLE COMPARTMENTS.

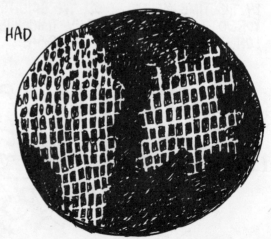

56

THE AVERAGE HUMAN CONTAINS
TRILLIONS OF CELLS.
OTHER CREATURES, SUCH
AS PROTOZOA,
CONSIST OF A SINGLE
CELL. CELLS COME IN
ALL SHAPES AND SIZES.

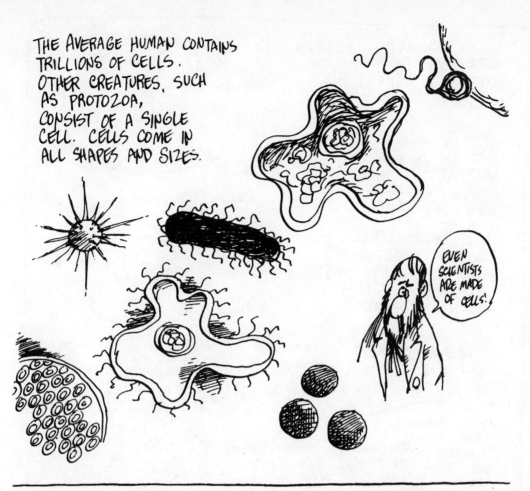

EVEN SCIENTISTS ARE MADE OF CELLS!

MOREOVER, SCIENTISTS SAW THAT ALL CELLS COME FROM THE
DIVISION OF A PRE-EXISTING CELL. BEFORE DIVISION,
EVERYTHING IN THE CELL IS DOUBLED.

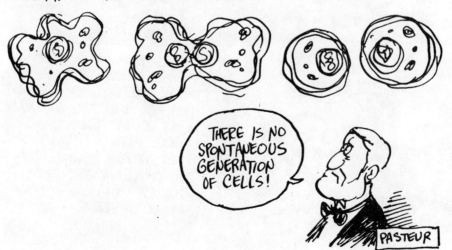

THERE IS NO SPONTANEOUS GENERATION OF CELLS!

PASTEUR

57

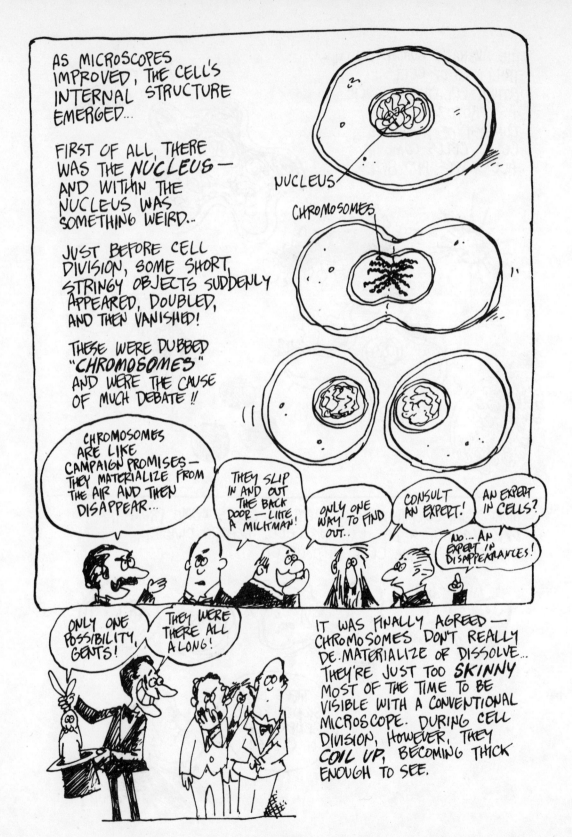

AS MICROSCOPES IMPROVED, THE CELL'S INTERNAL STRUCTURE EMERGED...

FIRST OF ALL, THERE WAS THE *NUCLEUS* — AND WITHIN THE NUCLEUS WAS SOMETHING WEIRD...

JUST BEFORE CELL DIVISION, SOME SHORT STRINGY OBJECTS SUDDENLY APPEARED, DOUBLED, AND THEN VANISHED!

THESE WERE DUBBED "*CHROMOSOMES*" AND WERE THE CAUSE OF MUCH DEBATE!!

NUCLEUS

CHROMOSOMES

CHROMOSOMES ARE LIKE CAMPAIGN PROMISES — THEY MATERIALIZE FROM THE AIR AND THEN DISAPPEAR...

THEY SLIP IN AND OUT THE BACK DOOR — LIKE A MILKMAN!

ONLY ONE WAY TO FIND OUT...

CONSULT AN EXPERT!

AN EXPERT IN CELLS?

NO... AN EXPERT IN DISAPPEARANCES!

ONLY ONE POSSIBILITY, GENTS!

THEY WERE THERE ALL ALONG!

IT WAS FINALLY AGREED — CHROMOSOMES DON'T REALLY DE-MATERIALIZE OR DISSOLVE... THEY'RE JUST TOO *SKINNY* MOST OF THE TIME TO BE VISIBLE WITH A CONVENTIONAL MICROSCOPE. DURING CELL DIVISION, HOWEVER, THEY *COIL UP*, BECOMING THICK ENOUGH TO SEE.

58

CAREFUL STUDY REVEALED WHAT HAPPENS TO CHROMOSOMES DURING CELL DIVISION.

FIRST—WHILE STILL INVISIBLE—THE CHROMOSOMES DUPLICATE THEMSELVES, REMAINING ATTACHED AT A SPOT CALLED THE CENTRO-MERE:

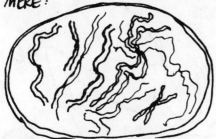

NEXT THEY THICKEN AND SHORTEN, BECOMING VISIBLE UNDER THE MICROSCOPE.

CENTRO-MERE

THE MEMBRANE AROUND THE NUCLEUS DISSOLVES, AND A FIBROUS *SPINDLE* FORMS, ON WHICH THE CHROMOSOMES LINE UP.

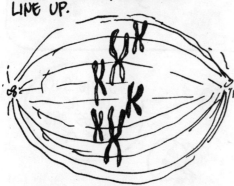

THE CENTROMERES DIVIDE AS THE SPINDLE FIBERS TUG THE CHROMOSOME PAIRS APART.

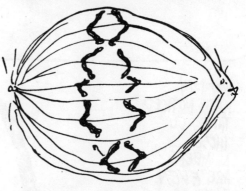

THE CHROMOSOMES ARRIVE AT THE OPPOSITE POLES, AND THE SPINDLE DISPERSES.

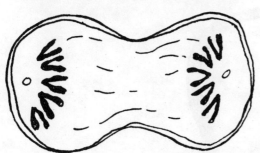

THE NUCLEAR MEMBRANE RE-FORMS; THE CHROMOSOMES UNWIND INTO INVISIBILITY; AND THE CELL DIVIDES.

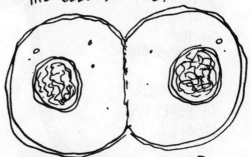

THIS PROCESS IS CALLED *MITOSIS.*

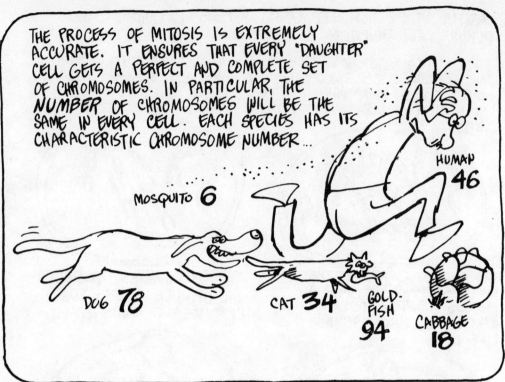

THE PROCESS OF MITOSIS IS EXTREMELY ACCURATE. IT ENSURES THAT EVERY "DAUGHTER" CELL GETS A PERFECT AND COMPLETE SET OF CHROMOSOMES. IN PARTICULAR, THE *NUMBER* OF CHROMOSOMES WILL BE THE SAME IN EVERY CELL. EACH SPECIES HAS ITS CHARACTERISTIC CHROMOSOME NUMBER...

MOSQUITO 6

HUMAN 46

DOG 78

CAT 34

GOLD-FISH 94

CABBAGE 18

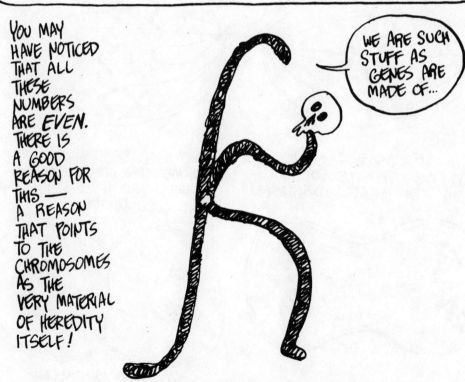

YOU MAY HAVE NOTICED THAT ALL THESE NUMBERS ARE *EVEN*. THERE IS A GOOD REASON FOR THIS — A REASON THAT POINTS TO THE CHROMOSOMES AS THE VERY MATERIAL OF HEREDITY ITSELF!

WE ARE SUCH STUFF AS GENES ARE MADE OF...

IT WAS THIS

FACT!

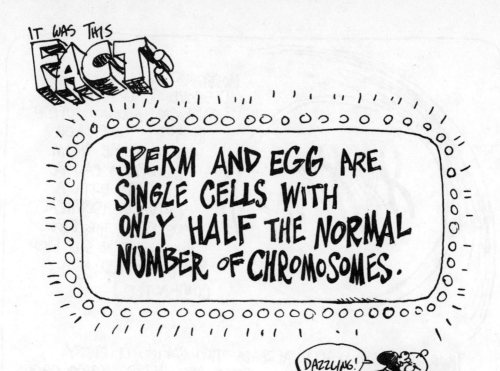

SPERM AND EGG ARE SINGLE CELLS WITH ONLY HALF THE NORMAL NUMBER OF CHROMOSOMES.

DAZZLING!

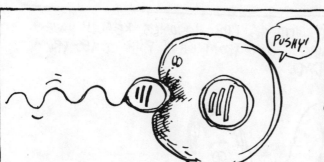

PUSHY!

IT WORKS LIKE THIS: THE SPERM AND EGG — THE *GERM CELLS*, OR *GAMETES*, AS THEY ARE KNOWN — EACH CARRIES A HALF SET OF CHROMOSOMES.

GLUP

AT FERTILIZATION, THEIR NUCLEI UNITE, GIVING THE FERTILIZED EGG, OR *ZYGOTE*, A FULL COMPLEMENT OF CHROMOSOMES. FROM THIS CELL ARISE ALL OTHERS BY MITOSIS.

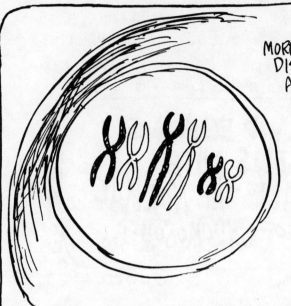

MOREOVER, IT WAS DISCOVERED (BY THE AMERICAN WILLIAM SUTTON IN 1902) THAT EACH CHROMOSOME FROM THE SPERM CAN BE MATCHED WITH A VIRTUALLY IDENTICAL ONE FROM THE EGG. (IT'S EASIER TO SEE WHEN THEY'RE DOUBLED AND CONTRACTED.)

THUS, THERE ARE REALLY ALREADY TWO COPIES OF EVERY CHROMOSOME IN THE CELL. THESE ARE CALLED "*HOMOLOGOUS PAIRS*"—"HOMOLOGOUS" MEANING "SAME SHAPE."

HUMANS, FOR EXAMPLE, WITH 46 CHROMOSOMES, REALLY HAVE 23* HOMOLOGOUS PAIRS: ONE FROM EACH PAIR COMES FROM MOM AND ONE FROM DAD.

THIS SUGGESTS THAT THERE MUST BE A SPECIAL KIND OF CELL DIVISION JUST FOR MAKING GAMETES...

*WITH ONE EXCEPTION, THE SEX CHROMOSOME. WE'LL EXPLAIN LATER!

THIS PROCESS, CALLED *MEIOSIS*, IS ACTUALLY A *DOUBLE* DIVISION:

AS IN MITOSIS, THE CHROMOSOMES DOUBLE AND THICKEN:

BUT THEN THE *HOMOLOGOUS* CHROMOSOMES PAIR OFF -- SOMEHOW!

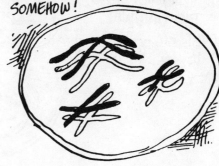

AGAIN THE SPINDLE FIBERS FORM AND THE CHROMOSOME QUARTETS ("TETRADS") LINE UP...

(MORE ON THIS LATER!)

AND THE *PAIRS* ARE SEPARATED. NOTE THE DIFFERENCE FROM MITOSIS!

WHEN THEY REACH THE POLES, THE SPINDLE VANISHES, AND *NEW* SPINDLES FORM "THE OTHER WAY."

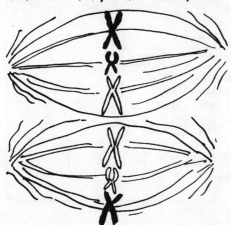

THE CHROMOSOMES THEN SEPARATE, AS IN MITOSIS.

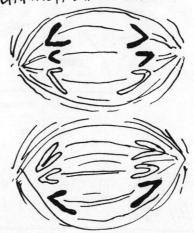

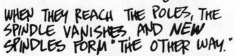

MEIOSIS RESULTS IN **FOUR** CELLS, EACH WITH **HALF** THE CHROMOSOMES OF THE ORIGINAL. COUNT 'EM — 3 VS. 6 IN THIS CASE.

BUT ALWAYS ONE FROM EACH HOMOLOGOUS PAIR!

NOTE THAT **WHICH** COPY ("HOMOLOG") OF EACH CHROMOSOME GOES TO WHICH CELL IS COMPLETELY RANDOM. EACH OF THESE COMBINATIONS IS EQUALLY AS LIKELY AS THE ONE ABOVE.

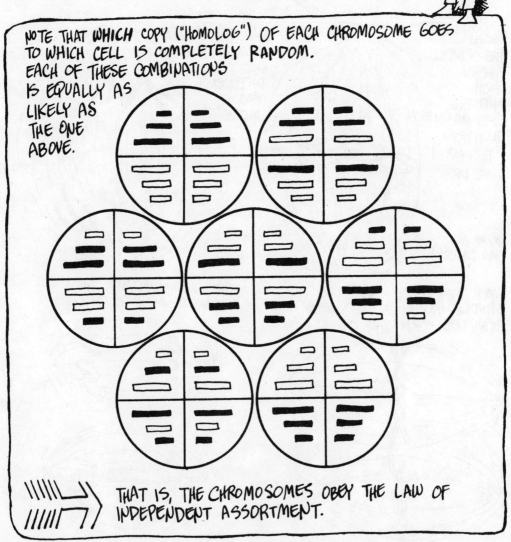

THAT IS, THE CHROMOSOMES OBEY THE LAW OF INDEPENDENT ASSORTMENT.

NCE MEIOSIS AND MITOSIS WERE UNDERSTOOD, BIOLOGISTS BEGAN TO SUSPECT THAT CHROMOSOMES MIGHT GOVERN HEREDITY... THEY LOOKED AGAIN AT PATTERNS OF INHERITANCE... AND SCIENCE AGAIN MARCHED — BACKWARD, TO THE LAWS OF MENDEL!!

FRIENDS OF YOURS?

TOWARD THE END OF THE 19TH CENTURY, THREE SCIENTISTS, WORKING INDEPENDENTLY, MORE OR LESS DUPLICATED THE AUSTRIAN MONK'S EXPERIMENTS AND RESULTS. THEY WERE:

HUGO DEVRIES

ERICH VON TSCHERMAK

CARL CORRENS

TO SUMMARIZE:

WHAT EXACTLY DID THEY REALIZE?

ANSWER:

CHROMOSOMES BEHAVE LIKE GENES. THEY RETAIN THEIR IDENTITY IN HYBRIDS, AND THEY SEGREGATE INDEPENDENTLY WHEN GERM CELLS ARE MADE. THEREFORE, IT'S LOGICAL TO ASSUME THAT GENES LIE ON CHROMOSOMES. (THERE MUST BE MANY GENES ON EACH ONE, BECAUSE THERE MUST BE FAR MORE GENES THAN THE FEW DOZEN CHROMOSOMES TYPICAL OF MOST SPECIES.)

A B c d E f ETC!

THE DISCOVERY OF HOMOLOGOUS PAIRS REALLY CINCHED THE CONNECTION TO MENDEL'S FINDINGS. REMEMBER, EACH CELL HAS A PAIR OF ALLELES FOR EACH GENE. NOW IT WAS REALIZED THAT:

THE TWO COPIES OF A GIVEN GENE LIE AT THE SAME POINT ON HOMOLOGOUS CHROMOSOMES.

I.E., IF ONE GENE FOR HEIGHT LIES HERE →

THEN THE OTHER COPY MUST BE HERE ←

ALL THIS TURNS OUT TO BE TRUE... BUT ONCE PEOPLE LOOKED MORE DEEPLY INTO THE MATTER, THEY DISCOVERED A FEW THINGS MENDEL HADN'T REALIZED...

FOR ONE THING, NOT ALL ORGANISMS HAVE A DOUBLE SET OF CHROMOSOMES. MANY LOWER SPECIES, LIKE SOME FUNGI, HAVE JUST A SINGLE SET.

LOWER THAN WHOM?

A CELL WITH A SINGLE SET OF CHROMOSOMES IS CALLED *HAPLOID*; ONE WITH TWO SETS IS CALLED *DIPLOID*. OUR BODY CELLS ARE DIPLOID, WHILE OUR GERM (SEX) CELLS ARE HAPLOID.

HAPLOID

DIPLOID

DIPLOID ORGANISMS INCLUDE ALL THE FAMILIAR MAMMALS AND BIRDS AND MANY PLANTS. HAPLOIDS INCLUDE MALE HONEY BEES, MANY FUNGI, AND ASEXUAL ONE-CELLED CREATURES.

BESIDES ALL THESE, THERE ARE ALSO *POLYPLOID* ORGANISMS, WITH MULTIPLE SETS OF CHROMOSOMES. A SURPRISING NUMBER OF EVERYDAY PLANTS ARE POLYPLOID. (NOT PEAS, THOUGH!!)

LIKE THE POTATO!

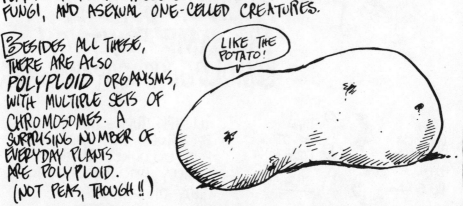

THE OTHER MAIN PROBLEM WITH MENDEL'S THEORY WAS THE PRINCIPLE OF INDEPENDENT ASSORTMENT. A PRECISE MEASURE OF HOW WRONG IT WAS LED TO THE ABILITY TO MAP OUT EXACTLY WHERE ON THE CHROMOSOME EACH OF ITS GENES MIGHT LIE... READ ON...

MAPMAKING

TO MENDEL—AND HIS HEIRS—GENES WERE JUST ABSTRACTIONS, LETTERS YOU COULD JUGGLE TO EXPLAIN AND PREDICT HOW HEREDITARY QUALITIES WOULD BE PASSED ALONG TO FUTURE GENERATIONS.

THEY'RE LIKE **GHOSTS**—INFLUENTIAL BUT INSUBSTANTIAL!

NOW IT APPEARED THAT GENES WERE ACTUAL, PHYSICAL OBJECTS. THEY LAY IN SOME ORDER ALONG THE CHROMOSOMES OF EVERY CELL, AND THE TWO ALLELES OF EACH GENE WERE ON THE TWO CHROMOSOMES OF A HOMOLOGOUS PAIR.

THEY'RE AS REAL AS BUMPS IN THE ROAD!

ONE MIGHT WONDER IF IT'S POSSIBLE TO MAKE A **GENE MAP** SHOWING JUST WHERE ON EACH CHROMOSOME ALL THESE HEREDITARY UNITS MIGHT LIE!!

69

THE ANSWER TO THIS DEPENDED ON A SEEMING *PARADOX*, FOR IN ONE RESPECT MENDEL'S FINDINGS CONFLICTED WITH THE OBSERVED BEHAVIOR OF CHROMOSOMES...

NAMELY— THE PRINCIPLE OF INDEPENDENT ASSORTMENT!

OBSERVE: THE NUMBER OF GENES MUST BE TREMENDOUS TO GOVERN A COMPLEX ORGANISM, BUT THE NUMBER OF CHROMOSOMES IN A CELL IS FAIRLY SMALL. A PEA PLANT HAS JUST *7 PAIRS* OF CHROMOSOMES, A HUMAN 23.

CONCLUSION: MANY GENES ON EACH CHROMOSOME!

THE PROBLEM: IF TWO GENES LIE ON THE SAME CHROMOSOME, *HOW CAN THEY BE INDEPENDENT??* AFTER ALL, CHROMOSOMES DON'T BREAK APART, DO THEY? SHOULDN'T DIFFERENT GENES SOMETIMES BE *LINKED??*

PHYSICALLY LINKED—BY THE CHROMOSOME!

SO—DO GENES ASSORT INDEPEN- DENTLY OR NOT?

WELL, IT TURNED OUT
TO BE SORT OF HALF-AND-HALF...

 THERE IS LINKAGE BETWEEN CERTAIN
GENES.... **BUT**

 CHROMOSOMES ALSO ENGAGE IN A
GOOD DEAL OF *GENE SWAPPING*,
OR (AS IT'S CALLED) *CROSSING OVER*.

TO ILLUSTRATE, LET'S LOOK
AT THE EXAMPLE OF THE
ORDINARY, GARDEN-VARIETY
TOMATO.

WITH MUTANT
MAYONNAISE?

...AND TRY NOT TO EAT
THE EXAMPLE UNTIL
AFTER CLASS...

TOMATOES HAVE A SKIN-TEXTURE GENE WITH A RECESSIVE ALLELE, *p*, WHICH CAUSES *HAIRY FRUIT.* (OF COURSE, YOU DON'T OFTEN SEE THESE IN THE MARKET !)

OO! A NEW GOURMET FRUIT!

NEW! FUZZY YUMMIES!

LIKEWISE, THE HEIGHT GENE HAS A RECESSIVE ALLELE, *d*, CAUSING DWARF PLANTS.

YOU LIKE IT?

I GIVE IT A *d*!

THE RESPECTIVE DOMINANT ALLELES ARE p^+, WHICH CAUSES SMOOTH FRUIT, AND d^+, WHICH MAKES TALL PLANTS.

TO TEST THE PRINCIPLE OF INDEPENDENT ASSORTMENT, WE CAN CROSS A DOUBLE RECESSIVE, *ppdd*, WITH A HETEROZYGOTE, pp^+dd^+.

HAIRY, DWARF
ppdd

SMOOTH, TALL
pp^+dd^+

SUPPOSE MENDEL WAS RIGHT, AND THE p's WERE INDEPENDENT OF THE d's.

THEN THE HYBRID pp⁺dd⁺ WOULD MAKE GAMETES WITH ALL COMBINATIONS OF p's AND d's.

pd   p⁺d

pd⁺   p⁺d⁺

CROSSING WITH THE DOUBLE RECESSIVE ppdd GIVES THIS:

	pd	p⁺d⁺	pd⁺	p⁺d
pd	pp⁺dd⁺	ppdd⁺	pp⁺dd	ppdd
pd	pp⁺dd⁺	ppdd⁺	pp⁺dd	ppdd
pd	pp⁺dd⁺	ppdd⁺	pp⁺dd	ppdd
pd	ppdd⁺	pp⁺dd	ppdd	

¼ SMOOTH, TALL   ¼ HAIRY, TALL   ¼ SMOOTH, DWARF   ¼ HAIRY, DWARF

NOW SUPPOSE p AND d LIE ON THE SAME CHROMOSOME. THEN THE HYBRID pp^+dd^+ HAS ITS ALLELES ON A HOMOLOGOUS PAIR:

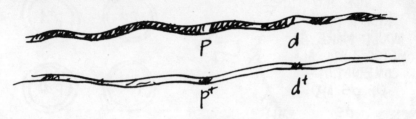

DURING MEIOSIS, THEY ARE SORTED OUT LIKE THIS:

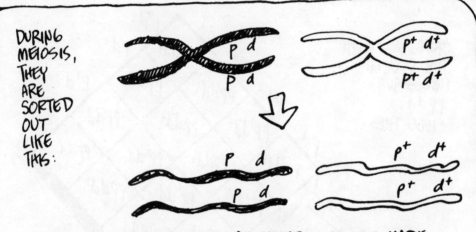

IN THIS CASE, ONLY TWO TYPES OF GAMETES CAN BE MADE: pd AND p^+d^+, RATHER THAN THE FOUR PREDICTED BY MENDEL.

CROSSING WITH THE DOUBLE RECESSIVE $ppdd$, WE GET

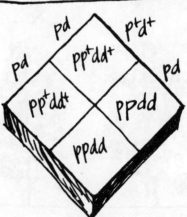

½ SMOOTH, TALL pp^+dd^+

½ HAIRY, DWARF $ppdd$

AND OF COURSE, WHO'S ON THE SIDE OF THE ANGELS?

WHEN THE CROSS IS ACTUALLY MADE, WHAT DOES ONE ACTUALLY GET: A 50-50 SPLIT OR AN EQUAL 4-WAY SPLIT?

IT SEEMS THAT *NEITHER* PREDICTION IS CORRECT. ALL FOUR TYPES DO APPEAR, BUT IN THESE PROPORTIONS:

SORRY, GREG!

SMOOTH, TALL
Pp^+dd^+
48%

HAIRY, TALL
$ppdd^+$
2%

SMOOTH, DWARF
Pp^+dd
2%

HAIRY, DWARF
$PP\ dd$
48%

IT'S O.K., I CAN TAKE THE DISAPPOINTMENT...

AFTER ALL, I AM DEAD — *SOB!*

IT'S CERTAINLY CLOSER TO THE PREDICTION BASED ON LINKAGE THAN TO MENDEL'S. BUT IF P AND d ARE LINKED, THEN WHERE DID THOSE 2% COMBINATIONS COME FROM ??

NOT TO PROLONG THE MYSTERY — THE GENES p AND d *ARE* ON THE SAME CHROMOSOME, BUT CHROMOSOMES CAN *EXCHANGE GENES.* IT'S CALLED *CROSSING OVER:*

DURING MEIOSIS, HOMOLOGUES LINE UP WITH CORRESPONDING ALLELES OPPOSITE ONE ANOTHER.

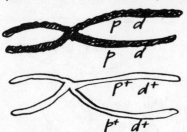

AT CERTAIN POINTS, SEEMINGLY "CHOSEN" AT RANDOM, THE CHROMOSOMES TOUCH:

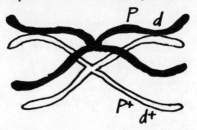

SOME SEGMENTS *CROSS OVER:*

WHEN THEY SEPARATE, THEY HAVE NEW COMBINATIONS OF ALLELES.

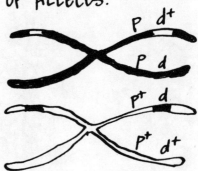

WHEN THAT HAPPENS TO OUR HETEROZYGOTE, SOME OF THE RESULTING GAMETES GET THE "RECOMBINANT" CHROMOSOMES. HENCE THE EXCEPTIONAL CROSSES!

} THE RECOM- BINANTS

NOTE: THANKS TO CROSSING OVER, THE CHROMOSOMES YOU PASS ALONG TO YOUR OFFSPRING ARE NOT EXACTLY YOUR OWN, BUT RATHER A SHUFFLED-TOGETHER COMBINATION!

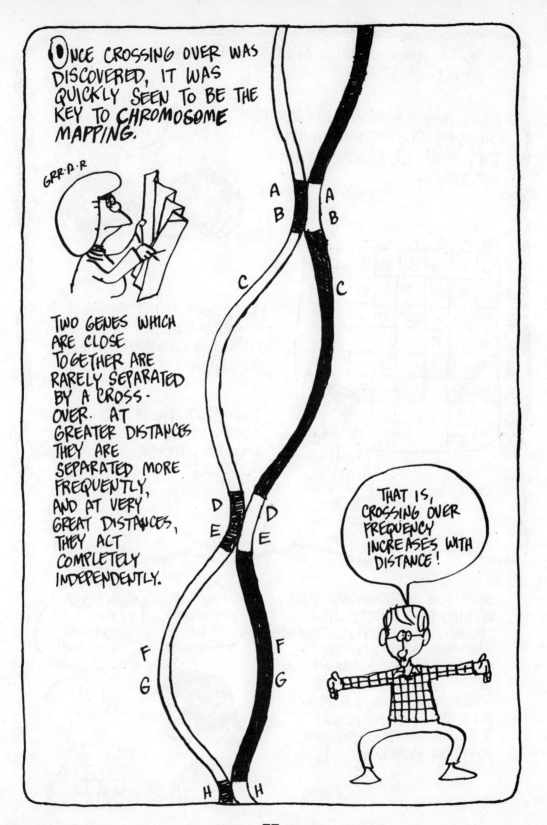

ONCE CROSSING OVER WAS DISCOVERED, IT WAS QUICKLY SEEN TO BE THE KEY TO *CHROMOSOME MAPPING.*

GRR·D·R

TWO GENES WHICH ARE CLOSE TOGETHER ARE RARELY SEPARATED BY A CROSS-OVER. AT GREATER DISTANCES THEY ARE SEPARATED MORE FREQUENTLY, AND AT VERY GREAT DISTANCES, THEY ACT COMPLETELY INDEPENDENTLY.

THAT IS, CROSSING OVER FREQUENCY INCREASES WITH DISTANCE!

77

So here's how you make a gene map without ever seeing a single gene:

First, make a vast number of crosses between individuals differing in various pairs of traits...

YOU'RE SOME TOMATO!

	A	B	C	D	E	F	G	H
A	0	.27	.03	.04	.33	.48	.19	.41
B	.27	0	.24	.31	.36	.45	.16	.44
C	.03	.24	0	.07	.30			
D	.04	.31	.07	0				
E	.33	.36	.30	.	0			
F	.48	.45				0		
G	.19	.16					0	
H	.41	.44						0

Next, see how often each pair is separated by crossing over (by looking at the offspring).

Then plot them out; those most closely linked will be closest together, etc!

I'VE BEEN MAPPED!

F   B D   A   C   G   E H

Since 1913, mapping has been applied to a variety of organisms. Nearly 1000 genes have been mapped in the bacterium E. COLI; about 300 in the tomato; 200 in the house mouse...; and a few hundred in human beings, although this was done by different means...

WHY THE DIFFERENCE FOR HUMANS?

THEY WON'T LET US DO BREEDING EXPERIMENTS..

78

MUTATION, or
A CHANGE OF GENES

WOW! A MUTATION IN PROGRESS!

SO FAR, WE HAVE BEEN THINKING OF GENES AS "ATOMS OF INHERITANCE"— UNCHANGING, *IMMUTABLE* UNITS OF HEREDITY.

A SLIGHT EXAGGERATION!

NOT ONLY ARE GENES *MUTABLE*, BUT THEY DO *MUTATE* FROM TIME TO TIME, OWING TO COPYING ERRORS AND VARIOUS ENVIRONMENTAL INFLUENCES.

THESE MUTATIONS —IT MEANS "CHANGES" IN LATIN— ARE FAIRLY RARE: THE CHANCE OF FINDING A MUTATION IN A GIVEN GENE IN AN INDIVIDUAL IS

→ 1 IN 100,000

THOUGH SOME GENES ARE MORE PRONE TO CHANGE THAN OTHERS!

EVEN AT THIS RATE, THEY DO ADD UP! A HUMAN HAS SOME 200,000 GENES, SO WE CARRY AN AVERAGE OF TWO NEW MUTATIONS APIECE.

GREAT SHADES!

SORRY... THOSE ARE MY EYES...

A CHANGE OF GENES MAY CAUSE GROSS DEFECTS, BUT MOST OF THE TIME, MUTATIONS ARE MORE SUBTLE. OTHERWISE, WE'D ALL LOOK LIKE MONSTERS.

SAY— YOU DON'T LOOK THAT GREAT!

SOMETIMES, MUTATIONS MERELY RESULT IN A NEW **RECESSIVE ALLELE**, LIKE HAIRINESS IN TOMATOES. YOU DON'T SEE ANYTHING AT ALL UNTIL TWO INDIVIDUALS WITH THE SAME MUTATION MATE TO FORM A HOMOZYGOTE. THEN ———

BLEAGH!

SOMETIMES MUTATIONS ARE COMPLETELY SILENT—PRODUCING NO CHANGE AT ALL—AND SOMETIMES THEY CAUSE CHANGES SO SLIGHT AS TO BE BARELY PERCEPTIBLE....

LIKE AN EXTRA MILLIMETER ON THE NOSE!

BUT EVERY SO OFTEN THE GENETIC "ERROR" MAY BE OF POSITIVE ADVANTAGE TO THE LUCKY MUTANT!!

HM! SO THE EGG *DID* COME BEFORE THE CHICKEN!

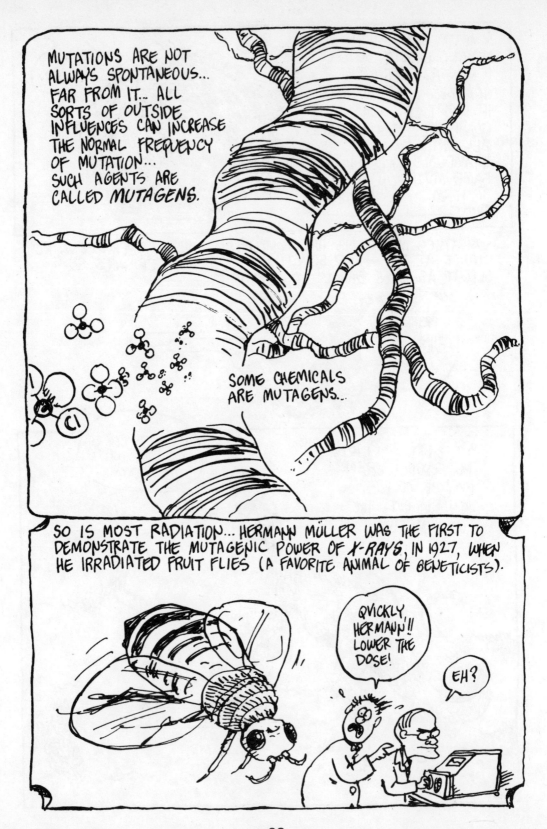

MUTATIONS ARE NOT ALWAYS SPONTANEOUS... FAR FROM IT... ALL SORTS OF OUTSIDE INFLUENCES CAN INCREASE THE NORMAL FREQUENCY OF MUTATION... SUCH AGENTS ARE CALLED *MUTAGENS.*

SOME CHEMICALS ARE MUTAGENS...

Cl

SO IS MOST RADIATION... HERMANN MÜLLER WAS THE FIRST TO DEMONSTRATE THE MUTAGENIC POWER OF *X-RAYS*, IN 1927, WHEN HE IRRADIATED FRUIT FLIES (A FAVORITE ANIMAL OF GENETICISTS).

QVICKLY HERMANN!! LOWER THE DOSE!

EH?

MUTATION IN BODY CELLS (*SOMATIC* CELLS, AS DISTINCT FROM GERM CELLS) MAY BE INVOLVED IN *CANCER*... IT MAKES SENSE: THE GENES CONTROL EVERYTHING ABOUT THE CELL, INCLUDING THE PROCESS OF DIVISION. ALTHOUGH THERE ARE STILL MANY MYSTERIES ABOUT CANCER, IT INVOLVES MUTATIONS THAT LEAD THE CELL TO DIVIDE OUT OF CONTROL.

MANY MUTAGENIC AGENTS ARE ALSO *CARCINOGENIC* (CANCER-CAUSING) — WHICH IS WHY THE FOOD + DRUG PEOPLE LOOK OUT FOR MUTAGENIC FOOD ADDITIVES... AND WHY YOU SHOULD LIMIT YOUR SUNBATHING, ESPECIALLY IF YOU HAVE PALE SKIN. (ULTRAVIOLET LIGHT IS MUTAGENIC.)

WHAT DETERMINES SEX?

THE COLOR OF PEA FLOWERS, THE TEXTURE OF TOMATOES, THE PINCHING OF PEA PODS — EACH OF THESE QUALITIES IS RULED BY A SINGLE GENE... BUT WHAT GOVERNS THAT MOST *OBVIOUS*, *INTERESTING*, AND (IN HUMANS) *SUBSTANTIAL* DIFFERENCE BETWEEN INDIVIDUALS: THE DIFFERENCE BETWEEN *MALE* AND *FEMALE*?

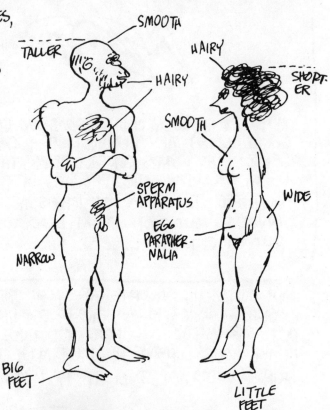

TALLER

SMOOTH

HAIRY

HAIRY

SMOOTH

SHORT-ER

SPERM APPARATUS

EGG PARAPHER-NALIA

NARROW

WIDE

BIG FEET

LITTLE FEET

THROUGHOUT HISTORY, SO MANY THINKERS TACKLED THIS QUESTION THAT ONE 18TH-CENTURY WRITER WAS INSPIRED TO COMPILE "262 GROUNDLESS HYPOTHESES." HIS OWN GROUNDLESS HYPOTHESIS BECAME THE 263RD....

IT'S THE HEAT OF THE MOON... OR THE BEAT OF MY TUNE...

OR THE BLEAT OF THE BABOON!

BUT OF COURSE IT'S
IN THE GENES...
NOT LONG AFTER
HOMOLOGOUS
CHROMOSOMES WERE
DISCOVERED,
SOMEBODY NOTICED
AN EXCEPTION:
HUMAN MALES HAVE
ONE PAIR THAT IS
NOT HOMOLOGOUS!!

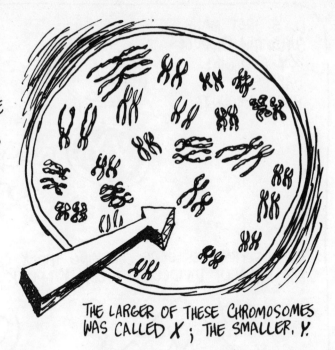

THE LARGER OF THESE CHROMOSOMES
WAS CALLED X ; THE SMALLER, Y.

THE ONLY GENETIC DIFFERENCE BETWEEN (HUMAN) MALES AND
FEMALES IS THIS:

FEMALES
HAVE
TWO
X
CHROMOSOMES:

WHILE
MALES
HAVE ONE
X AND
ONE Y:

THE OTHER 22 OTHER PAIRS OF CHROMOSOMES ARE THE SAME.

LET'S JUST MAKE SURE THIS PRODUCES BOY AND GIRL BABIES IN THE RIGHT AMOUNTS.

MEIOSIS PRODUCES EGGS CARRYING THE X CHROMOSOME; SPERM ARE EQUALLY DIVIDED BETWEEN X AND Y—

SO:

½ GIRLS

½ BOYS

WHAT A RELIEF!

HOWEVER, THE BASIC GENETIC QUESTION REMAINS: WHICH GENES ARE RESPONSIBLE FOR WHAT? IS IT THE Y CHROMOSOME THAT MAKES A MALE, OR DOES IT TAKE A DOUBLE DOSE OF X TO MAKE A FEMALE? WHAT WOULD HAPPEN TO SOMEBODY WITH *TWO X* CHROMOSOMES *AND A Y* ??

THIS ACTUALLY HAPPENS !!

THESE QUESTIONS ARE ANSWERED BY LOOKING AT CASES OF *FAULTY MEIOSIS*... SOMETIMES THERE IS AN ERROR IN MAKING SPERM:

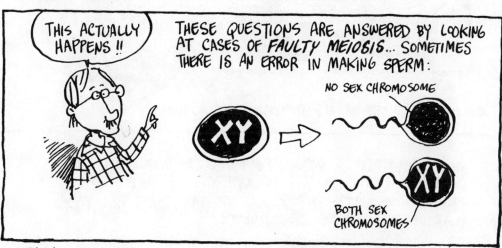

NO SEX CHROMOSOME

XY

BOTH SEX CHROMOSOMES

THEN:

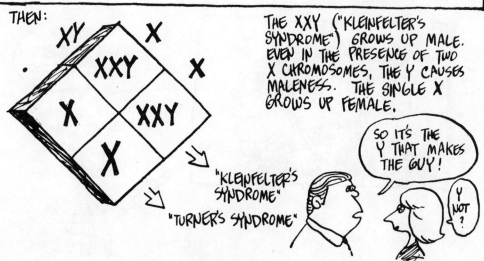

"KLEINFELTER'S SYNDROME"

"TURNER'S SYNDROME"

THE XXY ("KLEINFELTER'S SYNDROME") GROWS UP MALE. EVEN IN THE PRESENCE OF TWO X CHROMOSOMES, THE Y CAUSES MALENESS. THE SINGLE X GROWS UP FEMALE.

SO IT'S THE Y THAT MAKES THE GUY!

Y NOT?

ANOTHER ABNORMALITY IS THE "SUPER MALE" COMBINATION *XYY*, WHICH OCCURS IN ABOUT ONE BIRTH IN A THOUSAND. XYY CHILDREN GROW UP TO BE NORMAL MALES — EXCEPT THAT THEY END UP IN *PRISON* ABOUT *20 TIMES* MORE OFTEN THAN THE REST OF THE POPULATION. ABOUT 5% OF ALL PRISONERS HAVE AN EXTRA Y CHROMOSOME. SOME SAY:

*KARYOTYPE = AN ORGANISM'S PATTERN OF CHROMOSOMES

MOST GENETICISTS WOULD BE MORE CAUTIOUS... THE VAST MAJORITY (OVER 95%) OF XYY MALES ARE *NOT* IN PRISON... SO IT'S <u>IMPOSSIBLE</u> TO SAY THAT THE XYY KARYOTYPE CAUSES CRIMINALITY!

DO ANIMALS DO IT WITH X's AND Y's?

NOT NECESSARILY. SEX DETERMINATION IS HANDLED ALL SORTS OF WAYS, THOUGH MANY, MANY SPECIES HAVE THE SAME SYSTEM AS WE DO...

BUT AMONG BIRDS IT'S JUST THE *OPPOSITE* —

XX = MALE

XY = FEMALE

AND BEES ARE REALLY BEE-ZARRE: MALES DEVELOP FROM *UNFERTILIZED* EGGS. THEY'RE ALL *HAPLOID*, WHEREAS ALL *DIPLOIDS* ARE FEMALE (THE VAST MAJORITY OF THE HIVE). OTHERWISE, BEES HAVE NO SPECIFIC SEX CHROMOSOMES.

WILL YOU LISTEN TO ME? I SWEAR, BUSTER, IT'S LIKE YOU'RE ONLY HALF THERE SOMETIMES!

HUH?

THEN THERE ARE THE TRUE ODDITIES, WITH NO GENETIC DIFFERENCE BETWEEN MALE AND FEMALE AT ALL...

WHEN THE LARVA OF THE MARINE WORM *BONELLIA* SETTLES ON THE OCEAN FLOOR, IT DEVELOPS INTO A METER-LONG *FEMALE*.

BUT WHEN A LARVA LANDS ON A *FEMALE*, IT *WORMS* ITS WAY INTO HER BODY...

GLUP

...IN WHICH CASE, IT MATURES INTO A *MALE*, JUST A CENTIMETER LONG, AND PASSES ITS WHOLE LIFE INSIDE THE FEMALE!

WHICH WAY TO THE OVARIES?

AND SOMETIMES SEXUAL DIFFERENCES ARE SIMPLY SUBTLE... CERTAIN PROTOZOA HAVE TWO SEXES, BUT THEY DIFFER ONLY IN A SINGLE GENE... THESE ORGANISMS USUALLY REPRODUCE ASEXUALLY, AS FINDING AN APPROPRIATE PARTNER MUST NOT BE EASY!

EXCUSE ME— ARE YOU ANY DIFFERENT FROM ME?

IF YOU CAN'T TELL, HOW CAN I?

X-RATED GENES

NOW BACK TO HUMANS...
... WE'VE SEEN THAT ALL THE GENES ACCOUNTING FOR PURELY SEX-RELATED MATTERS HAVE ACCUMULATED ON JUST TWO CHROMOSOMES, X FOR FEMALE, Y FOR MALE...

WHERE ARE THE GENES FOR WINGS?

NOW WE MIGHT ASK THE FOLLOWING

QUESTION: ARE THERE ANY OTHER GENES ON THESE CHROMO- SOMES ????

THERE'S A GOOD REASON TO ASK: HUMANS EXHIBIT SEVERAL DEFECTS THAT APPEAR TO BE *SEX-LINKED*...

MOST **BALD** PEOPLE ARE **MEN**.

SO ARE MOST COLOR-BLIND PEOPLE.

DITTO FOR HEMO- PHILIACS.*

* HEMOPHILIA = A FAILURE OF THE BLOOD TO CLOT. HEMOPHILIACS CAN BLEED TO DEATH FROM A SMALL CUT.

91

FROM THIS YOU MIGHT CONCLUDE THAT THESE GENES LIE ON THE Y CHROMOSOME — BUT YOU'D BE WRONG!! ACTUALLY, HEMOPHILIA, COLOR-BLINDNESS, AND HEREDITARY BALDNESS ARE ALL CAUSED BY **RECESSIVE** ALLELES LYING ON THE X CHROMOSOME!!

I'M NOTHING BUT SEX!!

X Y

TAKE THE EXAMPLE OF BALDNESS:

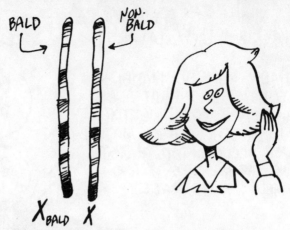

BALD → ← NON-BALD

X_BALD   X

THE REASON WOMEN ARE RARELY BALD IS THAT, EVEN IF THEY HAVE THE BALDNESS' ALLELE ON ONE X CHROMOSOME, THEY USUALLY HAVE THE DOMINANT *NON*-BALD ON THE OTHER.

BALD

X_BALD   Y

I CALL IT MY "RECESSIVE HAIRLINE!"

BUT IT SHOWS UP IN MEN BECAUSE THE Y CHROMOSOME HAS NO ALLELE FOR THAT GENE AT ALL. IN THE ABSENCE OF A DOMINANT ALLELE, THE RECESSIVE IS EXPRESSED!!

LET'S SEE HOW THESE *SEX-LINKED* GENES ARE PASSED ALONG:

SUPPOSE A NORMAL WOMAN (XX) HAS CHILDREN BY A BALD MAN ($X_{BALD}Y$).

THE DAUGHTERS ($X_{BALD}X$) ARE ALL CARRIERS.... NOT BALD THEMSELVES, THEY STILL CARRY THE RECESSIVE GENE. THE SONS ARE NORMAL.

IF YOUR MOTHER IS NORMAL, YOU CAN'T INHERIT BALDNESS FROM YOUR FATHER!

NEXT GENERATION: SUPPOSE ONE OF THE CARRIERS MARRIES A NORMAL MAN.

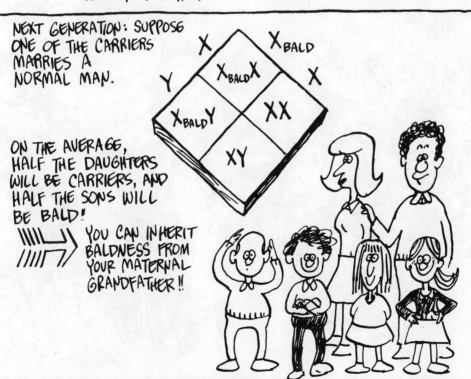

ON THE AVERAGE, HALF THE DAUGHTERS WILL BE CARRIERS, AND HALF THE SONS WILL BE BALD!

YOU CAN INHERIT BALDNESS FROM YOUR MATERNAL GRANDFATHER!!

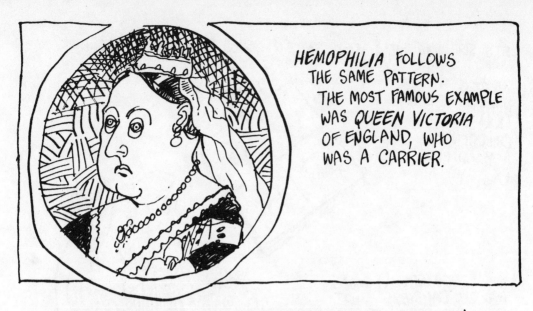

HEMOPHILIA FOLLOWS THE SAME PATTERN. THE MOST FAMOUS EXAMPLE WAS *QUEEN VICTORIA* OF ENGLAND, WHO WAS A CARRIER.

THERE IS NO RECORD OF HEMOPHILIA IN VICTORIA'S ANCESTORS, SO WE MAY ASSUME THE DEFECT APPEARED IN HER GENES AS A SPONTANEOUS MUTATION. THIS HAPPENS WITH HEMOPHILIA IN AN ESTIMATED 1 CASE IN EVERY 50,000 PARENTS.

HEMOPHILIA IS PASSED ALONG JUST LIKE BALDNESS, AND YOU CAN SEE THE PATTERN IN VICTORIA'S FAMILY TREE

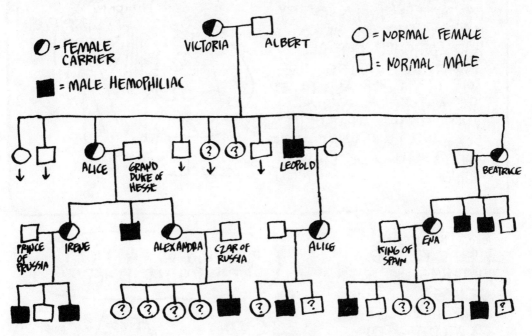

VICTORIA'S NUMEROUS BROOD INTERMARRIED WITH THE ROYAL HOUSES OF EUROPE, SPREADING THEIR MEDICAL PROBLEMS INTO PRUSSIA, SPAIN, AND PRE-REVOLUTIONARY RUSSIA...

WELL, JUST LOOK HOW FAR SCIENCE HAD COME BY THE EARLY 19TH CENTURY: MENDEL AND HIS HEIRS HAD POLISHED OFF ALL THOSE OLD PUZZLES: THE ROLE OF MOTHER AND FATHER, THE NATURE OF HYBRIDS AND "SPORTS," WHAT DETERMINES SEX, AND EVEN WHAT CAUSES THE QUALITIES OF LIVING THINGS...

ALL THESE HAD BEEN EXPLAINED IN TERMS OF **GENES**... GENES HAD BEEN LOCATED, MAPPED, AND THEIR PATTERNS OF INHERITANCE ANALYZED. NOW JUST ONE QUESTION REMAINED—

YES— WHY DO GENETICISTS WEAR POINTED BEARDS??

MUST BE GENETIC...

NO—THE QUESTION IS: WHAT **ARE** THE GENES, AND HOW DO THEY WORK??

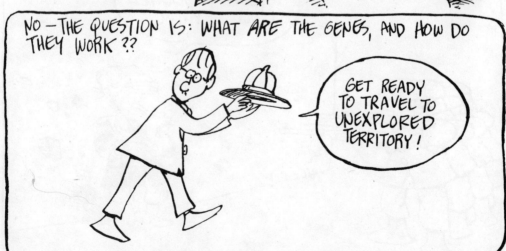

GET READY TO TRAVEL TO UNEXPLORED TERRITORY!

WHAT'S IN A CELL?

LET'S TAKE A CLOSER LOOK!

FROM HERE YOU CAN SEE HOW THE GORILLA IS MADE OF CELLS. TO UNDERSTAND THE APE, WE'LL HAVE TO MAKE SENSE OF WHAT GOES ON IN THESE TINY CHEMICAL FACTORIES.

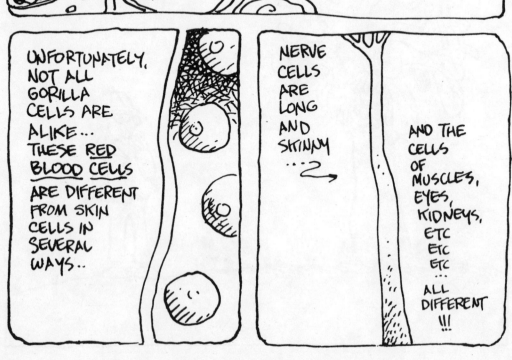

UNFORTUNATELY, NOT ALL GORILLA CELLS ARE ALIKE... THESE RED BLOOD CELLS ARE DIFFERENT FROM SKIN CELLS IN SEVERAL WAYS..

NERVE CELLS ARE LONG AND SKINNY ...

AND THE CELLS OF MUSCLES, EYES, KIDNEYS, ETC ETC ETC ... ALL DIFFERENT !!!

SIMILARLY, THE BANANA PRESENTS A WIDE DIVERSITY OF CELL TYPES...

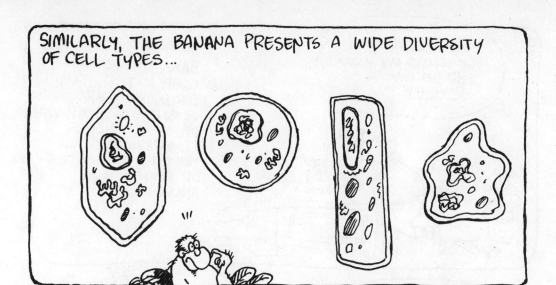

...EACH OF WHICH IS FILLED WITH ALL SORTS OF EVEN TINIER BODIES...

... MAKING BANANAS AND GORILLAS EXTREMELY HARD TO FIGURE OUT !!

HMM... THE GOLGI BODY CONNECTED TO THE ENDOPLASMIC RETICULUM... THE ENDOPLASMIC RETICULUM CONNECTED TO THE NUCLEAR MEMBRANE... NUCLEAR MEMBRANE CONNECTED TO... ¿SIGH¿

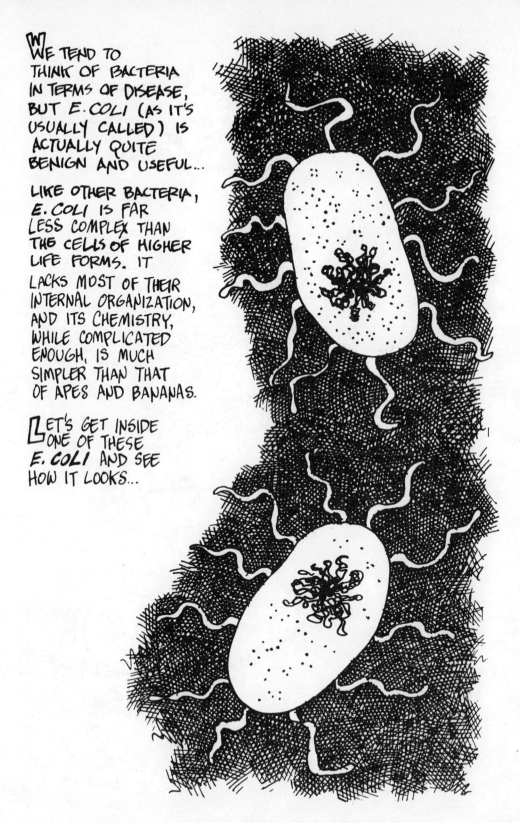

WE TEND TO THINK OF BACTERIA IN TERMS OF DISEASE, BUT *E. COLI* (AS IT'S USUALLY CALLED) IS ACTUALLY QUITE BENIGN AND USEFUL...

LIKE OTHER BACTERIA, *E. COLI* IS FAR LESS COMPLEX THAN THE CELLS OF HIGHER LIFE FORMS. IT LACKS MOST OF THEIR INTERNAL ORGANIZATION, AND ITS CHEMISTRY, WHILE COMPLICATED ENOUGH, IS MUCH SIMPLER THAN THAT OF APES AND BANANAS.

LET'S GET INSIDE ONE OF THESE *E. COLI* AND SEE HOW IT LOOKS...

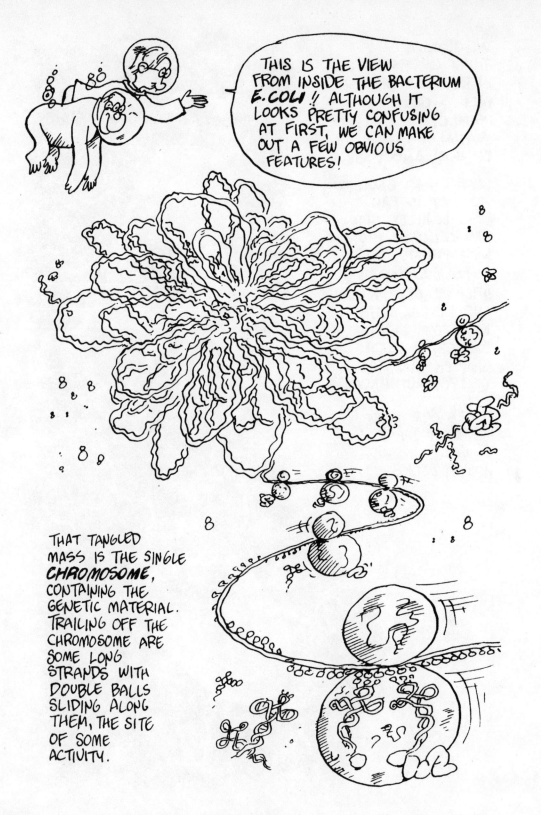

THIS IS THE VIEW FROM INSIDE THE BACTERIUM *E. COLI* !! ALTHOUGH IT LOOKS PRETTY CONFUSING AT FIRST, WE CAN MAKE OUT A FEW OBVIOUS FEATURES!

THAT TANGLED MASS IS THE SINGLE *CHROMOSOME*, CONTAINING THE GENETIC MATERIAL. TRAILING OFF THE CHROMOSOME ARE SOME LONG STRANDS WITH DOUBLE BALLS SLIDING ALONG THEM, THE SITE OF SOME ACTIVITY.

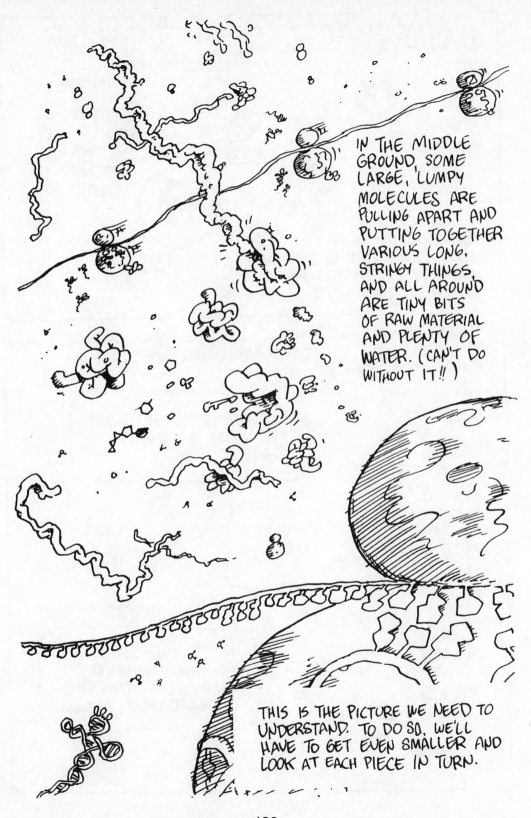

IN THE MIDDLE GROUND, SOME LARGE, LUMPY MOLECULES ARE PULLING APART AND PUTTING TOGETHER VARIOUS LONG, STRINGY THINGS, AND ALL AROUND ARE TINY BITS OF RAW MATERIAL AND PLENTY OF WATER. (CAN'T DO WITHOUT IT!!)

THIS IS THE PICTURE WE NEED TO UNDERSTAND. TO DO SO, WE'LL HAVE TO GET EVEN SMALLER AND LOOK AT EACH PIECE IN TURN.

MACROMOLECULES

(H) HYDROGEN

C CARBON

N NITROGEN

O OXYGEN

S SULFUR

P PHOSPHORUS

SURPRISING AS IT MAY SEEM, ALMOST EVERYTHING IN THAT COMPLEX PICTURE IS MADE OF JUST SIX DIFFERENT ELEMENTS.

IN THE CELL THESE ATOMS ARE JOINED TOGETHER TO FORM **MOLECULES**.

THE SIMPLEST AND MOST ABUNDANT BY FAR IS **WATER**, H_2O.

ANOTHER SMALL ONE IS THE PYRAMID-SHAPED **PHOSPHATE**, PO_4.

A BIT BIGGER ARE THE RING-SHAPED **SUGARS**. THIS ONE IS **GLUCOSE**, $C_6H_{12}O_6$.

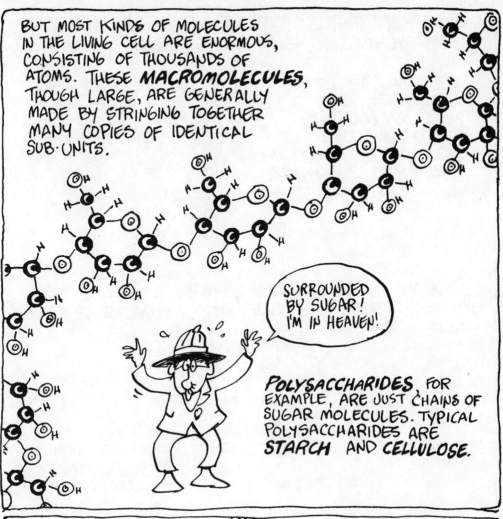

BUT MOST KINDS OF MOLECULES IN THE LIVING CELL ARE ENORMOUS, CONSISTING OF THOUSANDS OF ATOMS. THESE **MACROMOLECULES**, THOUGH LARGE, ARE GENERALLY MADE BY STRINGING TOGETHER MANY COPIES OF IDENTICAL SUB·UNITS.

SURROUNDED BY SUGAR! I'M IN HEAVEN!

POLYSACCHARIDES, FOR EXAMPLE, ARE JUST CHAINS OF SUGAR MOLECULES. TYPICAL POLYSACCHARIDES ARE **STARCH** AND **CELLULOSE.**

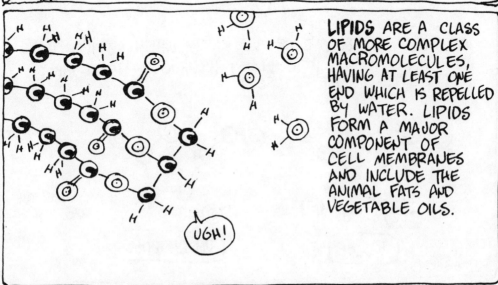

LIPIDS ARE A CLASS OF MORE COMPLEX MACROMOLECULES, HAVING AT LEAST ONE END WHICH IS REPELLED BY WATER. LIPIDS FORM A MAJOR COMPONENT OF CELL MEMBRANES AND INCLUDE THE ANIMAL FATS AND VEGETABLE OILS.

UGH!

STILL MORE COMPLEX, BUT MOST IMPORTANT IN GENETICS, ARE THE *NUCLEIC ACIDS* AND *PROTEINS*... WATCH CLOSELY:

THE BUILDING BLOCKS FOR NUCLEIC ACIDS ARE CALLED *NUCLEOTIDES.* AN INDIVIDUAL NUCLEOTIDE ITSELF HAS 3 COMPONENTS: A *SUGAR*, A *PHOSPHATE*, AND A *BASE*, LIKE SO —

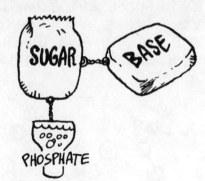

THESE ARE HOOKED TOGETHER TO MAKE A LONNNNNNNG SUGAR-PHOSPHATE "BACKBONE" WITH A SEQUENCE OF BASES STICKING OFF:

sugar — base
/
phosphate
|
sugar — base
/
phosphate
|
sugar — base
/
phosphate
|
ETC !!

THIS MAY GO ON FOR *MILLIONS* OF NUCLEOTIDES!

THE SUGAR MAY BE ONE OF TWO KINDS, WHICH WE ILLUSTRATE HERE WITHOUT ALL THEIR PESKY HYDROGEN ATOMS. (THEY JUST CLUTTER UP THE PICTURE!)

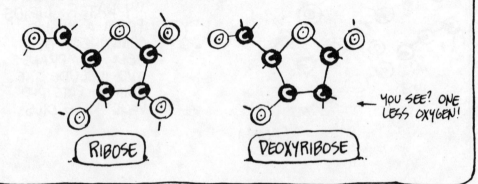

RIBOSE

DEOXYRIBOSE

← YOU SEE? ONE LESS OXYGEN!

THE PHOSPHATE 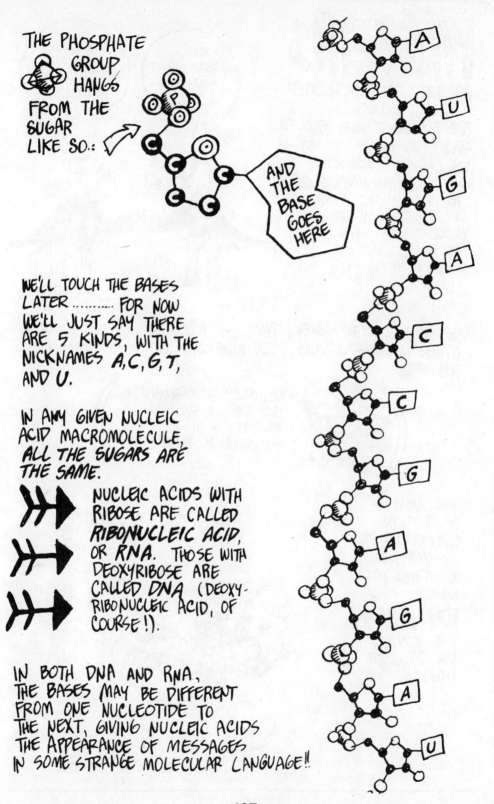 GROUP HANGS FROM THE SUGAR LIKE SO.:

P O O O
C C C O C C C

AND THE BASE GOES HERE

WE'LL TOUCH THE BASES LATER .... FOR NOW WE'LL JUST SAY THERE ARE 5 KINDS, WITH THE NICKNAMES A, C, G, T, AND U.

IN ANY GIVEN NUCLEIC ACID MACROMOLECULE, ALL THE SUGARS ARE THE SAME.

➤➤➤ NUCLEIC ACIDS WITH RIBOSE ARE CALLED *RIBONUCLEIC ACID,* OR *RNA.* THOSE WITH DEOXYRIBOSE ARE CALLED *DNA* (DEOXY-RIBONUCLEIC ACID, OF COURSE!).

IN BOTH DNA AND RNA, THE BASES MAY BE DIFFERENT FROM ONE NUCLEOTIDE TO THE NEXT, GIVING NUCLEIC ACIDS THE APPEARANCE OF MESSAGES IN SOME STRANGE MOLECULAR LANGUAGE!!

A
U
G
A
C
C
G
A
G
A
U

PROTEINS

ARE THE MOST COMPLICATED MACROMOLECULES OF ALL. THE BIOLOGIST *MAX PERUTZ* SPENT 25 YEARS — MOST OF HIS CAREER — ANALYZING JUST **ONE** OF THEM: *HEMOGLOBIN,* THE PROTEIN THAT CARRIES OXYGEN THROUGH THE BLOODSTREAM. FOR THIS, PERUTZ RECEIVED THE NOBEL PRIZE IN 1962...

MAX, WHAT TOOK YOU SO LONG?

IT WOULD TAKE TOO MUCH TIME TO EXPLAIN...

YET IN A CERTAIN SENSE, PROTEINS ARE SIMPLE, TOO: LIKE OTHER MACROMOLECULES, THEY ARE LONG CHAINS OF SMALLER SUBUNITS.

ACTUALLY, HEMOGLOBIN IS TWO PAIR OF SUCH CHAINS, WRAPPED UP IN A SYMMETRICAL TANGLE.

THE SUBUNITS OF PROTEIN MOLECULES ARE *AMINO ACIDS,* WHICH ARE NOT NAMED AFTER **IDI AMIN,** THE FORMER DICTATOR OF UGANDA.

ABSURD!! OF COURSE THEY ARE!

TAKE IT FROM ME!

OH! O.K...

THE TYPICAL AMINO ACID
LOOKS LIKE THIS:

IT'S THAT CLUSTER OF
"OTHER ATOMS" THAT
COMPLICATES MATTERS...

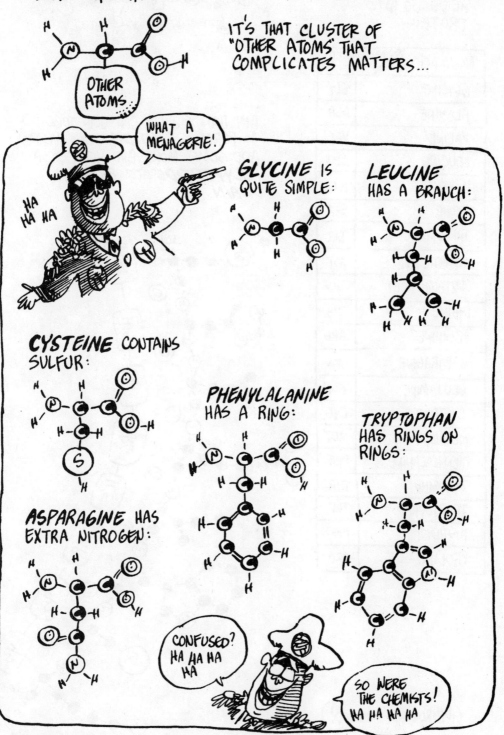

IN ALL, SOME 20
"STANDARD" AMINO
ACIDS GO INTO
PROTEINS:

AMINO ACID	ABBREVIATED AS:
GLYCINE	GLY
ALANINE	ALA
VALINE	VAL
LEUCINE	LEU
ISOLEUCINE	ILE
SERINE	SER
THREONINE	THR
ASPARTIC ACID	ASP
GLUTAMIC ACID	GLU
LYSINE	LYS
ARGININE	ARG
ASPARAGINE	ASN
GLUTAMINE	GLN
CYSTEINE	CYS
METHIONINE	MET
PHENYLALANINE	PHE
TYROSINE	TYR
TRYPTOPHAN	TRP
HISTIDINE	HIS
PROLINE	PRO

(HYDROGENS OMITTED!)

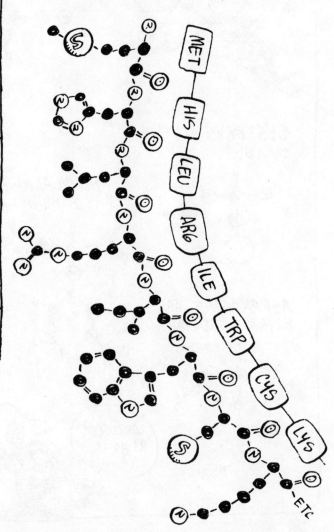

ANY TWO OF THEM CAN JOIN TOGETHER
TO FORM A *PEPTIDE*...VERY PEPPY...
ADD SOME MORE AND YOU GET A
POLYPEPTIDE, OR *PROTEIN
CHAIN*...

110

NOT ONE ACID GETS OUT OF LINE!!

CHUCKLE CHORTLE

EVERY PROTEIN HAS A PRECISE NUMBER AND SEQUENCE OF AMINO ACIDS. MUTUAL ATTRACTIONS AMONG THEM CAUSE THE CHAIN TO COIL UP INTO A FAIRLY COMPACT, BUT FLEXIBLE SHAPE.

(OFTEN, AS WITH HEMOGLOBIN, SEVERAL POLYPEPTIDE CHAINS MAY COIL TOGETHER.)

WHAT DO PROTEINS DO FOR A CELL? YOU PROBABLY THINK OF THEM AS SOMETHING THAT ENRICHES SHAMPOO... OR MAYBE YOU KNOW ABOUT THE PROTEIN IN FINGERNAILS, FEATHERS, AND HAIR... BUT ACTUALLY, MOST PROTEINS ARE SOMETHING ELSE AGAIN...

NEW! PROTEIN-ENRICHED GLORP AVOID HAIRY STOMACH— DO NOT DRINK!

MOST PROTEINS ARE ENZYMES!!

HA HA HA OUCH

ENZYMES ARE PROTEINS WHICH TAKE APART OR PUT TOGETHER OTHER MOLECULES. EACH ENZYME IS RESPONSIBLE FOR JUST ONE SPECIFIC REACTION.

A TYPICAL ENZYME LIES IN WAIT FOR THE RIGHT MOLECULES TO COME AROUND.

THE ENZYME BINDS TO THE SMALL MOLECULES...

...AND COMBINES THEM...

...INTO A NEW MOLECULE, WHICH IS RELEASED.

THE ENZYME ITSELF REMAINS UNCHANGED IN THE PROCESS.

IN A SIMILAR WAY, *DIGESTIVE* ENZYMES BREAK DOWN LARGE MOLECULES. SEVERAL KINDS, FOR EXAMPLE, CHOP SUGARS OFF POLYSACCHARIDES !!

THESE PROTEINS ARE SO IMPORTANT BECAUSE VIRTUALLY EVERY ONE OF LIFE'S CHEMICAL REACTIONS IS DRIVEN BY SOME ENZYME.

WHEN CHEMICALS COME UP THROUGH THE ROOTS OF THE BANANA TREE, THE PLANT'S ENZYMES CONVERT THEM INTO THE CONSTITUENTS OF A BANANA...

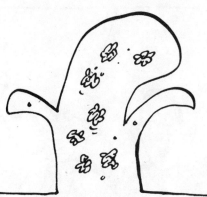

MMMM

GURGLE

THEN, WHEN THE GORILLA EATS THE BANANA, THE APE'S ENZYMES DIGEST THE FRUIT AND TURN IT INTO AN APE...

...AND LIKEWISE FOR *E. COLI*, WHICH HAS ITS OWN ENZYMES...!

IN OTHER WORDS:

An organism is made by its enzymes.

AND WHAT DO YOU SUP-POSE MAKES THE ENZYMES?

ONE GENE, ONE ENZYME

BEADLE

TATUM

THE RELATIONSHIP BETWEEN GENES AND ENZYMES FIRST BECAME CLEAR IN THE 1940's, THANKS TO EXPERIMENTS PERFORMED BY BIOLOGISTS **GEORGE BEADLE** AND **EDWARD TATUM**, WORKING WITH MUTANT STRAINS OF THE COMMON BREAD MOLD **NEUROSPORA** GROWN IN BATHS OF CHEMICAL NUTRIENTS.

EACH MUTANT WAS FOUND TO REQUIRE **MORE CHEMICAL NUTRIENTS** IN ITS DIET THAN WERE NEEDED BY NORMAL MOLD FOR EXAMPLE, ONE MUTANT HAD TO BE FED AN EXTRA AMINO ACID, WHILE ANOTHER REQUIRED A CERTAIN VITAMIN.

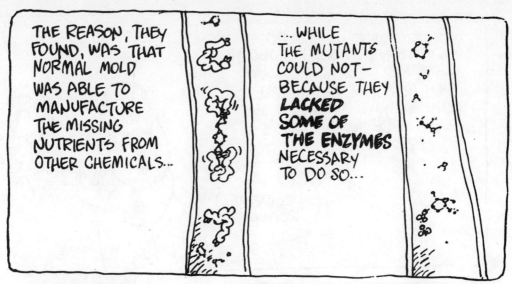

THE REASON, THEY FOUND, WAS THAT NORMAL MOLD WAS ABLE TO MANUFACTURE THE MISSING NUTRIENTS FROM OTHER CHEMICALS...

...WHILE THE MUTANTS COULD NOT— BECAUSE THEY **LACKED SOME OF THE ENZYMES** NECESSARY TO DO SO...

BY EXHAUSTIVE CROSS-BREEDING AND BIOCHEMICAL ANALYSIS, THE SCIENTISTS DISCOVERED THIS: **THE MUTATION OF A SINGLE GENE LED TO THE LACK OF A SINGLE ENZYME** ...

OR, TO TURN IT AROUND...

* *

The metabolic role of the genes is to make enzymes, and each gene is responsible for one, specific enzyme.

IN SHORT: **ONE GENE, ONE ENZYME !!**

115

SO THAT'S WHAT GENES *DO*—MAKE ENZYMES—BUT STILL NOBODY UNDERSTOOD EXACTLY WHAT THEY *WERE*... THOUGH A FIRST STEP IN THAT DIRECTION HAD BEEN MADE IN THE 1920's BY *FRED GRIFFITH*...

BY ACCIDENT, REALLY!

GRIFFITH WORKED WITH TWO STRAINS OF THE PNEUMONIA BACTERIUM *PNEUMOCOCCUS.* ONE WAS THE VIRULENT "WILD TYPE" FOUND IN NATURE.

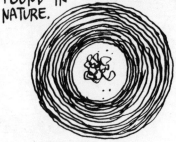

THE OTHER LACKED A CERTAIN ENZYME USED IN MAKING THE THICK OUTER CAPSULE SEEN IN THE WILD TYPE.

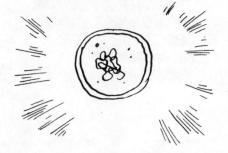

WHEN INJECTED INTO MICE, THE WILD TYPE INVARIABLY CAUSED DISEASE...

THE MUTANT *PNEUMO-COCCUS,* ON THE OTHER HAND, HAD NO EFFECT.

WHEW!

NOW GRIFFITH BOILED SOME OF THE WILD TYPE, MANGLING AND KILLING THEM.

AS EXPECTED, THESE HEAT-KILLED BACTERIA DID NO HARM.

ARE YOU KIDDING? MY NERVES ARE SHOT!!

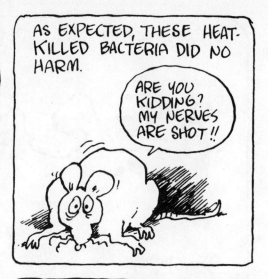

THEN, JUST TO BE THOROUGH, GRIFFITH MIXED SOME *HEAT-KILLED WILD TYPE* WITH *LIVE MUTANTS.*

DESPITE THE FACT THAT EACH INGREDIENT WAS HARMLESS IN ITSELF—

?

NOT ONLY DID THE MICE DIE, BUT *LIVE* WILD-TYPE *PNEUMOCOCCUS* WERE FOUND IN THEIR BODIES! GRIFFITH COULDN'T FIGURE THIS OUT AT ALL !!!

EVENTUALLY, IT WAS UNDERSTOOD THIS WAY:

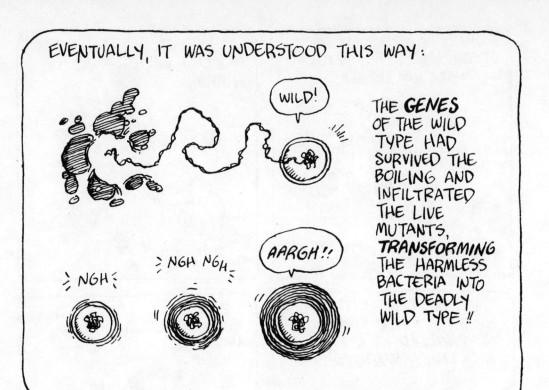

WILD!

NGH

NGH NGH

AARGH!!

THE **GENES** OF THE WILD TYPE HAD SURVIVED THE BOILING AND INFILTRATED THE LIVE MUTANTS, **TRANSFORMING** THE HARMLESS BACTERIA INTO THE DEADLY WILD TYPE!!

IN THE 1940's, *OSWALD AVERY* SET OUT TO IDENTIFY THIS "TRANSFORMING FACTOR:"

BOILING BACTERIA BY THE VATFUL, AVERY PRECIPITATED, EXTRACTED, CENTRIFUGED, ANALYZED, OVER AND OVER...

UNTIL HE HAD A THIMBLEFUL OF PURE GENETIC MATERIAL...

118

WHEN AVERY ANNOUNCED HIS RESULTS IN 1940, FEW SCIENTISTS BELIEVED HIM!!

THE SPIRAL STAIRCASE

DNA? IS THAT A GOVERNMENT AGENCY?

BEFORE AVERY, SCIENTISTS HAD PAID LITTLE ATTENTION TO D N A.

THEY KNEW IT CONTAINED THE SUGAR *DEOXYRIBOSE*, PLENTY OF *PHOSPHATE*, AND FOUR *BASES*.

THE FOUR BASES ARE KNOWN AS **A, C, G,** AND **T**, WHICH ARE SHORT FOR:

ADENINE

CYTOSINE

GUANINE

THYMINE

THESE WERE ASSUMED TO BE PRESENT IN EQUAL PROPORTIONS.

AFTER AVERY, HOWEVER, RESEARCHERS BEGAN TO LOOK MORE CLOSELY...

ERWIN CHARGAFF FOUND.

① THE COMPOSITION OF DNA VARIED FROM ONE SPECIES TO ANOTHER, IN PARTICULAR IN THE RELATIVE AMOUNTS OF THE BASES A, C, T, G.

② IN ANY DNA, *THE NUMBER OF A's* WAS *THE SAME AS THE NUMBER OF T's*; SIMILARLY, THE NUMBER OF C's WAS EQUAL TO THE NUMBER OF G's.

WHAT DID THIS MEAN? CHARGAFF COULDN'T SAY...

BY STUDYING X-RAY PICTURES OF DNA, *ROSALIND FRANKLIN* WAS ABLE TO SHOW THAT THE DNA MOLECULE PROBABLY HAD THE CORKSCREW SHAPE OF A *HELIX* WITH TWO OR THREE CHAINS...

BUT WAS IT TWO OR THREE...?

IN 1952 **JAMES WATSON** AND **FRANCIS CRICK** CRACKED THE PUZZLE.

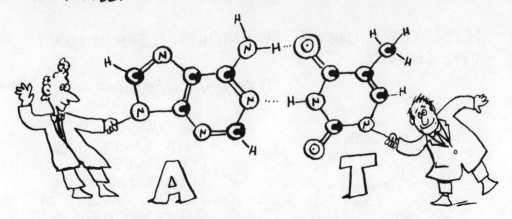

BY PLAYING WITH SCALE-MODEL ATOMS, THEY OBSERVED THAT **ADENINE** FITTED TOGETHER WITH **THYMINE**, WHILE **GUANINE** PAIRED NATURALLY WITH **CYTOSINE**.

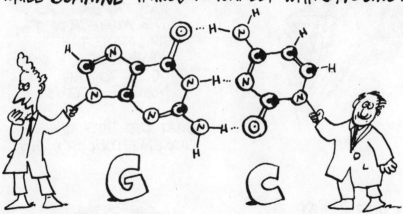

EACH BASE PAIR WOULD BE HELD TOGETHER BY **HYDROGEN** BONDING, A WEAK ATTRACTION THAT MAY OCCUR BETWEEN A HYDROGEN ON ONE MOLECULE AND A NON-HYDROGEN ATOM ON ANOTHER MOLECULE.

IT WAS ALSO CLEAR **A** DID NOT FIT WITH **C**, NOR **G** WITH **T**.

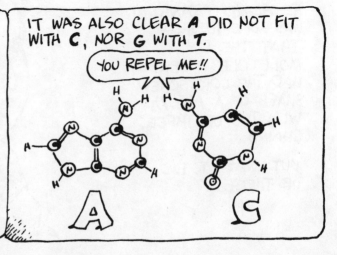

EACH OF THESE TWO **BASE PAIRS** IS NEARLY FLAT:

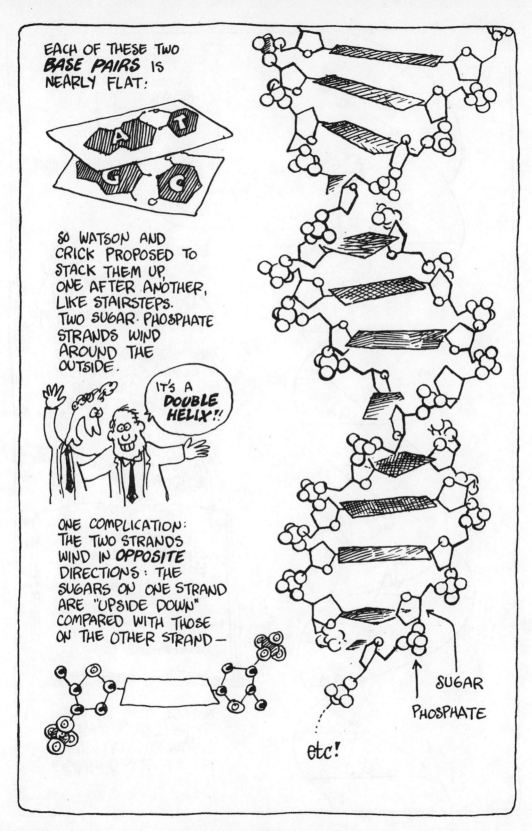

SO WATSON AND CRICK PROPOSED TO STACK THEM UP, ONE AFTER ANOTHER, LIKE STAIRSTEPS. TWO SUGAR·PHOSPHATE STRANDS WIND AROUND THE OUTSIDE.

IT'S A **DOUBLE HELIX**!!

ONE COMPLICATION: THE TWO STRANDS WIND IN **OPPOSITE** DIRECTIONS: THE SUGARS ON ONE STRAND ARE "UPSIDE DOWN" COMPARED WITH THOSE ON THE OTHER STRAND—

SUGAR

PHOSPHATE

etc!

123

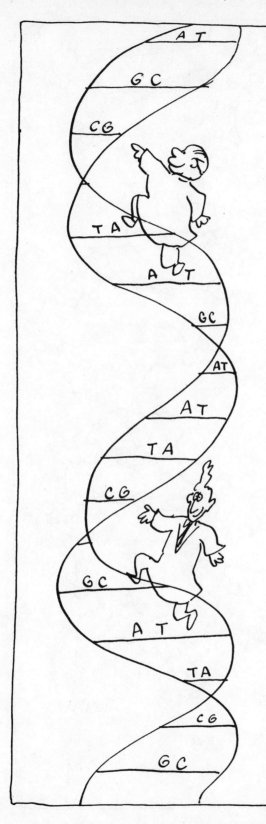

THIS MODEL CLEARLY EXPLAIN'S *CHARGAFF'S* OBSERVATION THAT THE NUMBER OF *T's* IS EQUAL TO THE NUMBER OF *A's*: *T* AND *A* ARE ALWAYS PAIRED TOGETHER!

DITTO FOR *G* AND *C*!

THIS IS THE *PRINCIPLE OF COMPLEMENTARITY*: EACH BASE CAN PAIR WITH ONLY ONE OTHER, CALLED ITS *COMPLEMENT*.

*W*ATSON AND CRICK GOT THE IDEA!! THEY WROTE:

"It has not escaped our notice that the pairing... immediately suggests a possible copying mechanism for the genetic material."

IN FACT, IT IS THE KEY TO THE GENE'S MAIN FUNCTIONS: REPLICATION AND PROTEIN SYNTHESIS.

REPLICATION

GENE-COPYING, OR DNA *REPLICATION*, AS WATSON AND CRICK SAW, IS SIMPLE IN PRINCIPLE. *EACH STRAND OF THE DOUBLE HELIX CONTAINS THE INFORMATION NECESSARY TO MAKE ITS COMPLEMENTARY STRAND.*

SCHEMATICALLY, IT WORKS LIKE THIS: WHEN THE DNA IS READY TO MULTIPLY, ITS TWO STRANDS PULL APART:

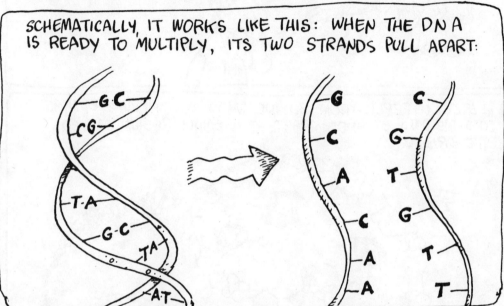

ALONG EACH ONE, A NEW STRAND FORMS IN THE ONLY POSSIBLE WAY:

WE WIND UP WITH TWO COPIES OF THE ORIGINAL!

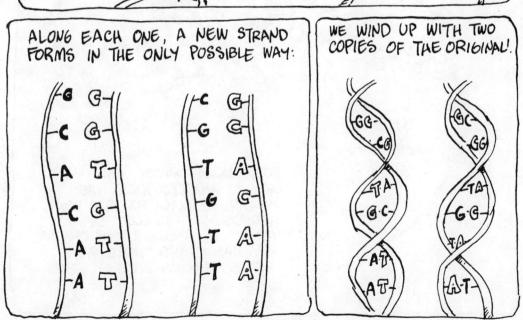

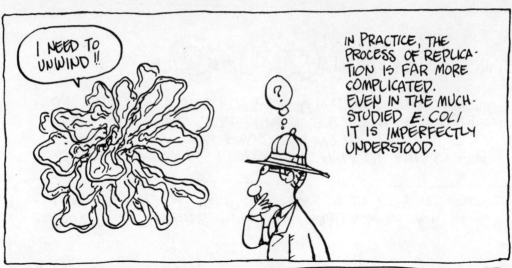

I NEED TO UNWIND !!

IN PRACTICE, THE PROCESS OF REPLICATION IS FAR MORE COMPLICATED. EVEN IN THE MUCH-STUDIED *E. COLI* IT IS IMPERFECTLY UNDERSTOOD.

IN *E. COLI* REPLICATION BEGINS WHEN A "SNIPPING" ENZYME CUTS THE DNA STRANDS APART AT A SMALL REGION CALLED THE **ORIGIN**.

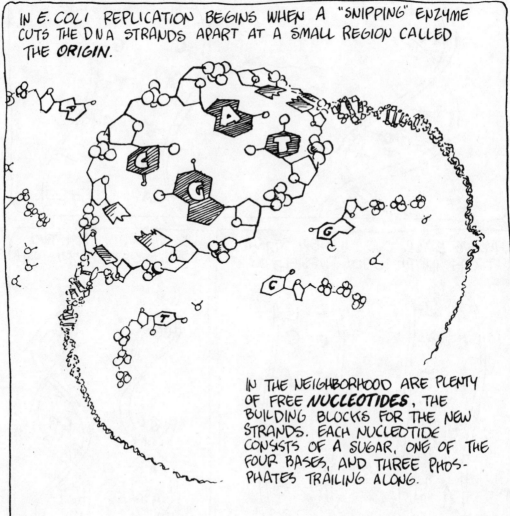

IN THE NEIGHBORHOOD ARE PLENTY OF FREE **NUCLEOTIDES**, THE BUILDING BLOCKS FOR THE NEW STRANDS. EACH NUCLEOTIDE CONSISTS OF A SUGAR, ONE OF THE FOUR BASES, AND THREE PHOSPHATES TRAILING ALONG.

WHEN A FREE NUCLEOTIDE MEETS ITS COMPLEMENTARY BASE ON THE D.N.A., IT STICKS, WHILE THE "WRONG" NUCLEOTIDES BOUNCE AWAY.

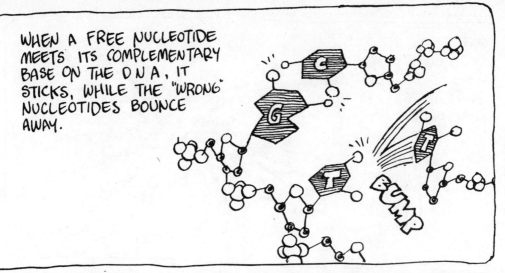

AS THE "SNIPPING" ENZYME OPENS THE D.N.A. FURTHER, MORE NUCLEOTIDES ARE ADDED, AND A "CLIPPING" ENZYME PUTS THEM TOGETHER, KNOCKING OFF THE EXTRA PHOSPHATES.

THIS PROCEEDS ALONG BOTH STRANDS SIMULTANEOUSLY — IN OPPOSITE DIRECTIONS. THE "CLIPPING" ENZYME CAN GO ONLY ONE WAY, RUNNING SMOOTHLY DOWN ONE STRAND, WHILE BACKING UP THE OTHER IN A SERIES OF SPURTS.

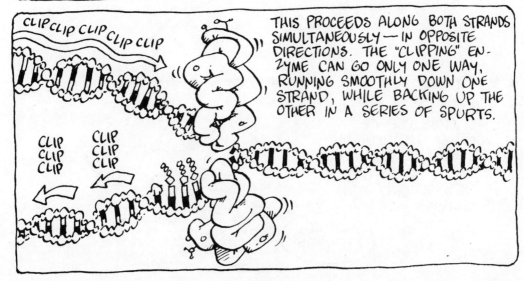

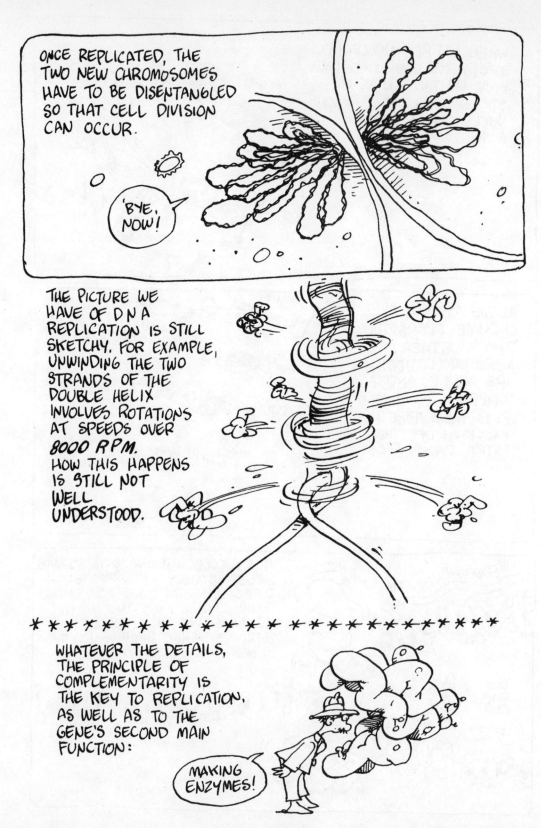

ONCE REPLICATED, THE TWO NEW CHROMOSOMES HAVE TO BE DISENTANGLED SO THAT CELL DIVISION CAN OCCUR.

'BYE, NOW!

THE PICTURE WE HAVE OF D N A REPLICATION IS STILL SKETCHY. FOR EXAMPLE, UNWINDING THE TWO STRANDS OF THE DOUBLE HELIX INVOLVES ROTATIONS AT SPEEDS OVER *8000 RPM.* HOW THIS HAPPENS IS STILL NOT WELL UNDERSTOOD.

* *

WHATEVER THE DETAILS, THE PRINCIPLE OF COMPLEMENTARITY IS THE KEY TO REPLICATION, AS WELL AS TO THE GENE'S SECOND MAIN FUNCTION:

MAKING ENZYMES!

The MOLECULE is the MESSAGE

Enzymes and other proteins come in many shapes, but in an important respect, they are all alike.

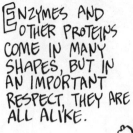

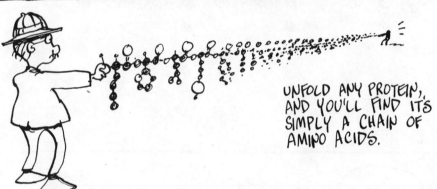

Unfold any protein, and you'll find it's simply a chain of amino acids.

Let go of the ends, and the protein will re-fold itself, owing to the mutual attractions among the components.

(Actually, many proteins need help from another "chaperone" protein to fold up.)

That is: THE SEQUENCE DETERMINES THE STRUCTURE.

IN VIEW OF THE RELATIONSHIP BETWEEN GENES AND PROTEINS, THIS SUGGESTS THAT THE *SEQUENCE* OF D.N.A MUST SOMEHOW PARALLEL OR REFLECT THE *SEQUENCE* OF THE PROTEIN.

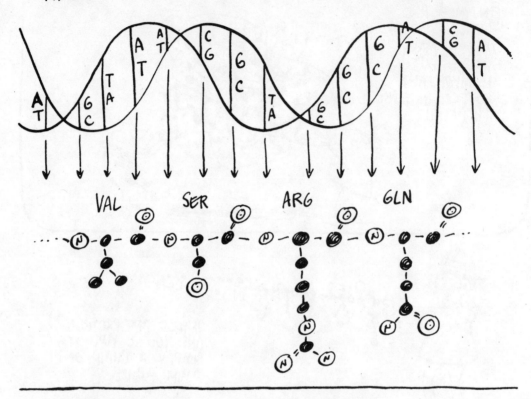

THE MAIN IDEA:

The sequence of base pairs may be thought of as a series of "words" specifying the order of amino acids in each protein.

WHY DOES LIFE HAVE TO BE SO COMPLICATED?

TO MAKE THE TRANSLATION FROM DNA "WORDS" TO AMINO ACIDS, SOME SOPHISTICATED MOLECULAR MACHINERY COMES INTO PLAY...

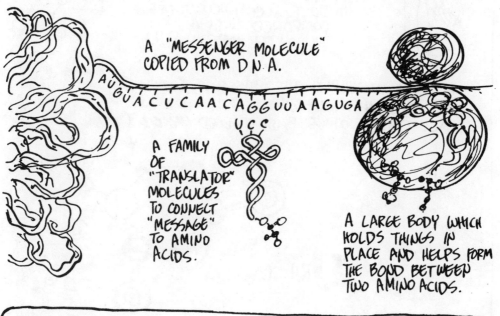

A "MESSENGER MOLECULE" COPIED FROM D.N.A.

A U G U A C U C A A C A G G U U A A G U G A
U C C

A FAMILY OF "TRANSLATOR" MOLECULES TO CONNECT "MESSAGE" TO AMINO ACIDS.

A LARGE BODY WHICH HOLDS THINGS IN PLACE AND HELPS FORM THE BOND BETWEEN TWO AMINO ACIDS.

ALL THREE OF THESE AGENTS ARE PARTLY OR ENTIRELY MADE OF THAT *OTHER* NUCLEIC ACID:

R-N-A-A-A-AY.

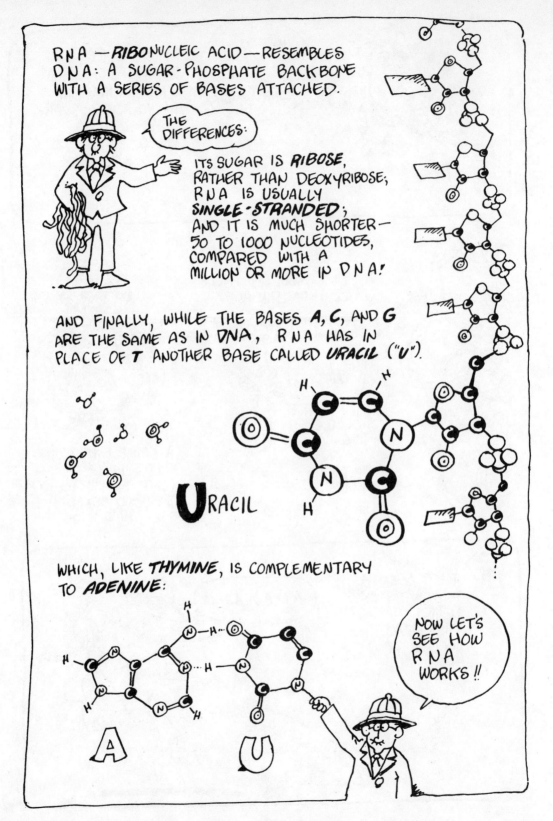

RNA —**RIBO**NUCLEIC ACID— RESEMBLES DNA: A SUGAR-PHOSPHATE BACKBONE WITH A SERIES OF BASES ATTACHED.

THE DIFFERENCES:

ITS SUGAR IS **RIBOSE**, RATHER THAN DEOXYRIBOSE; RNA IS USUALLY **SINGLE-STRANDED**; AND IT IS MUCH SHORTER— 50 TO 1000 NUCLEOTIDES, COMPARED WITH A MILLION OR MORE IN DNA!

AND FINALLY, WHILE THE BASES **A, C,** AND **G** ARE THE SAME AS IN **DNA**, RNA HAS IN PLACE OF **T** ANOTHER BASE CALLED **URACIL** ("U").

URACIL

WHICH, LIKE **THYMINE**, IS COMPLEMENTARY TO **ADENINE**:

NOW LET'S SEE HOW RNA WORKS!!

A   **U**

PROTEIN SYNTHESIS BEGINS WHEN A REGION OF DNA IS TEASED APART AND A MOLECULE OF RNA IS BUILT ALONG ONE STRAND BY AN ENZYME CALLED *RNA POLYMERASE*. THIS PROCESS IS CALLED *TRANSCRIPTION*.

IT HAPPENS AS IN DNA REPLICATION: EACH BASE OF THE RNA IS COMPLEMENTARY TO THE CORRESPONDING BASE ON THE DNA.

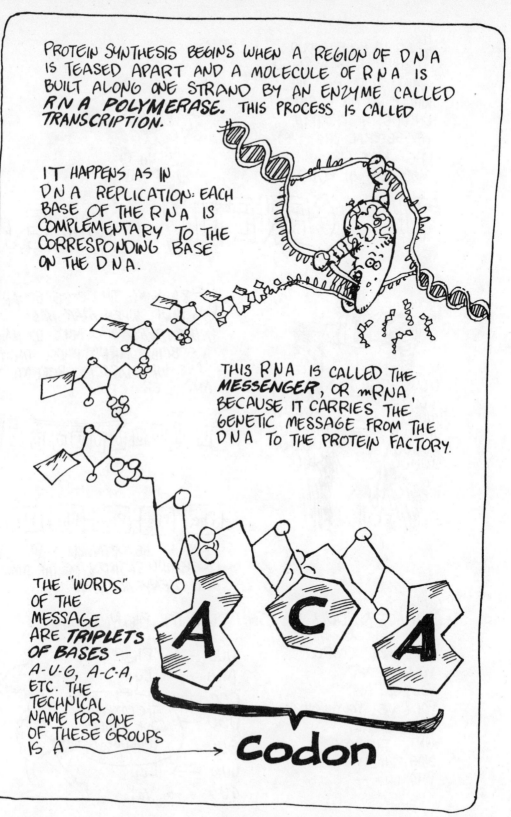

THIS RNA IS CALLED THE *MESSENGER*, OR mRNA, BECAUSE IT CARRIES THE GENETIC MESSAGE FROM THE DNA TO THE PROTEIN FACTORY.

THE "WORDS" OF THE MESSAGE ARE *TRIPLETS OF BASES* — A-U-G, A-C-A, ETC. THE TECHNICAL NAME FOR ONE OF THESE GROUPS IS A ⟶ codon

Each 3-BASE CODON STANDS FOR A SINGLE AMINO ACID, AND THE WHOLE mRNA STRAND ENCODES A PROTEIN (OR SEVERAL PROTEINS). IT'S JUST LIKE A MESSAGE IN CODE—

THE GENETIC CODE!

CRACKING THIS CODE BEGAN IN 1961, WHEN *MARSHALL NIRENBERG* WAS ABLE TO MAKE A SPECIAL mRNA, WHOSE ONLY BASE WAS *URACIL*, REPEATED OVER AND OVER: "POLY-U."

From it he obtained a protein consisting entirely of the amino acid *PHENYLALANINE*.

SO ☞ UUU WAS THE CODON FOR PHENYLALANINE...

NEXT THEY DECODED POLY-A, AND POLY-C, AND POLY-UG, POLY-UGU, ETC, ETC, ETC, UNTIL THE CODE WAS FINALLY BROKEN—

UUU → Phe
AAA → Lys
CCC →
UGU →
GUU →

THE COMPLETE CODE TABLE FOLLOWS!

UUG → Leu
GUG → Val

134

	U	C	A	G	
U	UUU } PHE UUC UUA } LEU UUG	UCU UCC UCA } SER UCG	UAU } TYR UAC UAA } STOP UAG	UGU } CYS UGC UGA STOP UGG TRP	U C A G
C	CUU CUC } LEU CUA CUG	CCU CCC } PRO CCA CCG	CAU } HIS CAC CAA } GLN CAG	CGU CGC } ARG CGA CGG	U C A G
A	AUU AUC } ILE AUA AUG MET	ACU ACC } THR ACA ACG	AAU } ASN AAC AAA } LYS AAG	AGU AGC } SER AGA } ARG AGG	U C A G
G	GUU GUC } VAL GUA GUG	GCU GCC } ALA GCA GCG	GAU } ASP GAC GAA } GLU GAG	GGU GGC } GLY GGA GGG	U C A G

FIRST LETTER

THIRD LETTER

SOME SALIENT POINTS:

THE CODE IS **REDUNDANT**: WITH 64 POSSIBLE CODONS, BUT ONLY 20 AMINO ACIDS, THERE MUST BE "SYNONYMS," DIFFERENT CODONS WHICH ENCODE THE SAME AMINO ACID.

THERE ARE "STOP" SIGNALS. THREE CODONS DO NOT ENCODE ANY AMINO ACID AT ALL. THESE SERVE TO TERMINATE MESSAGES.

ALSO: THE CODE IS NON-OVERLAPPING. THE "WORDS" FOLLOW EACH OTHER WITHOUT GAPS OR OVERLAPS. WE'LL SEE SHORTLY HOW IT KNOWS WHERE TO START...

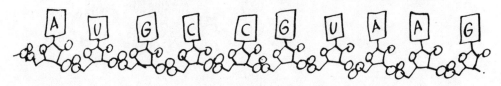

135

THE ACTUAL TRANSLATORS
OF THE GENETIC
CODE ARE A GROUP
OF RNA MOLECULES
CALLED *TRANSFER* RNA,
OR tRNA. OWING TO
PAIRING AMONG ITS
BASES, tRNA'S
TWIST THEMSELVES
INTO THIS KEY SHAPE.

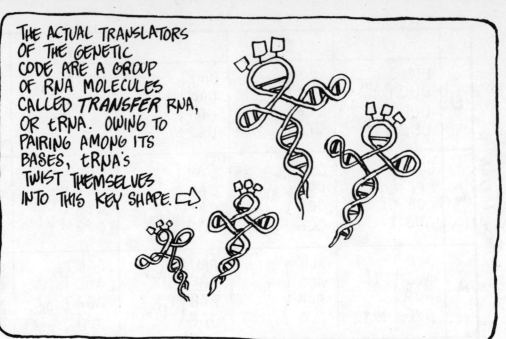

THE LOOP
END OF tRNA
HAS THREE
UNPAIRED BASES.
THIS "*ANTICODON*"
MAY BIND
WITH THE
COMPLEMENTARY
CODON OF
mRNA.
AT THE "TAIL"
END OF
tRNA IS A
SITE FOR
ATTACHING
A SINGLE
AMINO ACID.

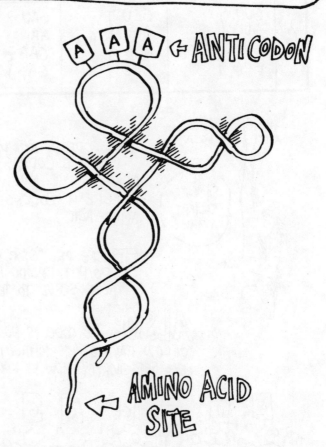

A A A ← ANTICODON

← AMINO ACID SITE

FOR EACH ANTICODON, THERE IS AN ENZYME WHICH RECOGNIZES IT AND ATTACHES THE APPROPRIATE AMINO ACID TO ITS tRNA.

ONCE THEY ARE LINKED, THE ANTICODON BINDS TO THE COMPLEMENTARY CODON OF MESSAGE.

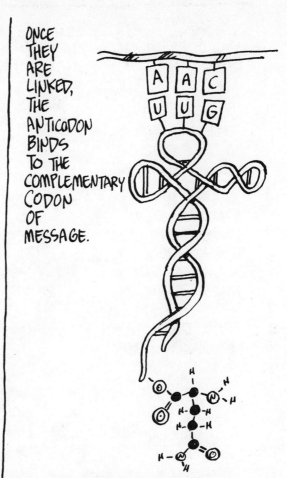

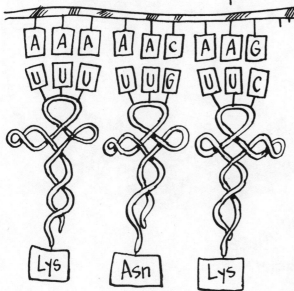

SCHEMATICALLY, THIS IS THE WAY A STRING OF BASES IS TRANSLATED INTO A SEQUENCE OF AMINO ACIDS. **HOWEVER,** THE CELL NEEDS ONE MORE PIECE OF EQUIPMENT TO MAKE IT WORK: THE *RIBOSOME.*

137

HOW PROTEINS ARE MADE

THE FINAL INGREDIENT IN THE PROTEIN-MAKING APPARATUS IS AN OBJECT THAT HOLDS EVERYTHING IN PLACE.

THIS IS THE **RIBOSOME**, A DOUBLE BALL OF ABOUT 50 PROTEINS WRAPPED UP WITH R N A. THIS R N A IS CALLED **RIBOSOMAL RNA**, rRNA FOR SHORT.

THE RIBOSOME HAS TWO SLOTS IN WHICH MOLECULES OF tRNA CAN FIT SNUGLY.

NOW TO MAKE A PROTEIN:
WHEN THE mRNA READS
OUT THE DNA SEQUENCE,
IT ENTERS A SEA
OF RIBOSOMES.

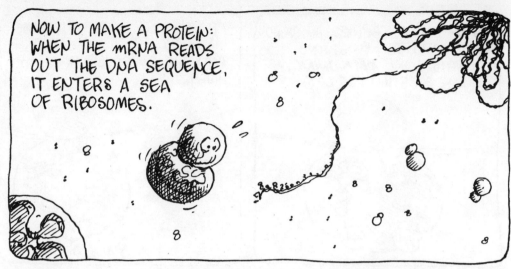

ONE HALF AT A TIME, A RIBO-
SOME BINDS ONTO THE mRNA.

CHOMP

WHUMP

THE BINDING SITE IS
LOCATED AT OR NEAR THE
CODON A·U·G.

A U G

THUS, A·U·G IS ALWAYS THE
FIRST "WORD" OF EVERY MESSAGE.

AUGAAA

A·U·G AND THE NEXT
CODON EACH BOND WITH
COMPLEMENTARY tRNA's,
WHICH FIT INTO THE SLOTS
ON THE RIBOSOME.

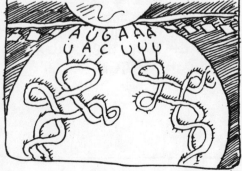

AUGAAA
UAC

EACH tRNA CARRIES AN AMINO ACID (AA), THE FIRST ONE ALWAYS BEING *METHIONINE*, WHICH GOES WITH *A·U·G.*

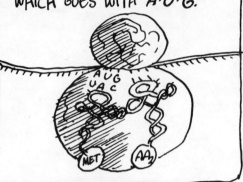

AN ENZYME IN THE RIBOSOME LINKS THE TWO AMINO ACIDS, AND THE FIRST tRNA FLOATS AWAY.

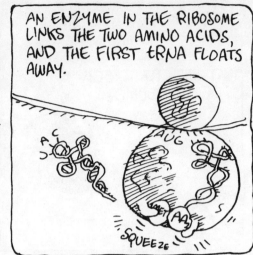

THE RIBOSOME THEN MOVES DOWN THREE MORE BASES.

ANOTHER tRNA AND AMINO ACID BIND ON.

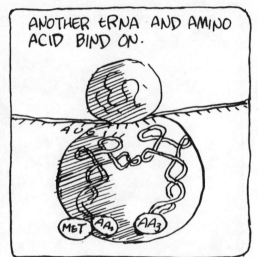

THE AMINO ACIDS ARE LINKED; THE "EMPTY" tRNA IS DISCARDED; AND SO THE RIBOSOME MOVES ALONG THE MESSAGE, PILING UP AMINO ACIDS, WHICH FOLD THEMSELVES INTO A PROTEIN.

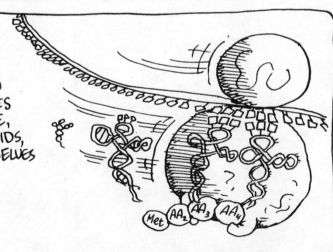

THIS PROCESS CONTINUES UNTIL THE RIBOSOME REACHES ONE OF THE "STOP" SIGNALS.

IT STOPS BECAUSE THERE IS NO tRNA WITH AN ANTICODON TO MATCH.

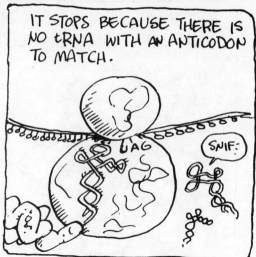

SNIF!

THE COMPLETED PROTEIN IS CLIPPED OFF BY ANOTHER RIBOSOMAL ENZYME.

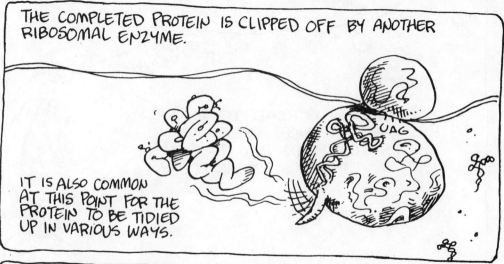

IT IS ALSO COMMON AT THIS POINT FOR THE PROTEIN TO BE TIDIED UP IN VARIOUS WAYS.

FINALLY, THE RIBOSOME, MESSAGE, AND tRNA DISSOCIATE...

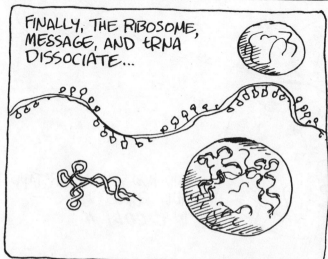

...AND THE NEW MACROMOLECULE GOES OFF TO DO ITS JOB: STRUCTURE, ENZYME, OR WHATEVER...

CHOMP

SQUEEZE

GRIND

IN THE LIVING CELL, ALL THESE PROCESSES ARE GOING ON TOGETHER. THIS IS HOW IT LOOKS IN *E. COLI.*

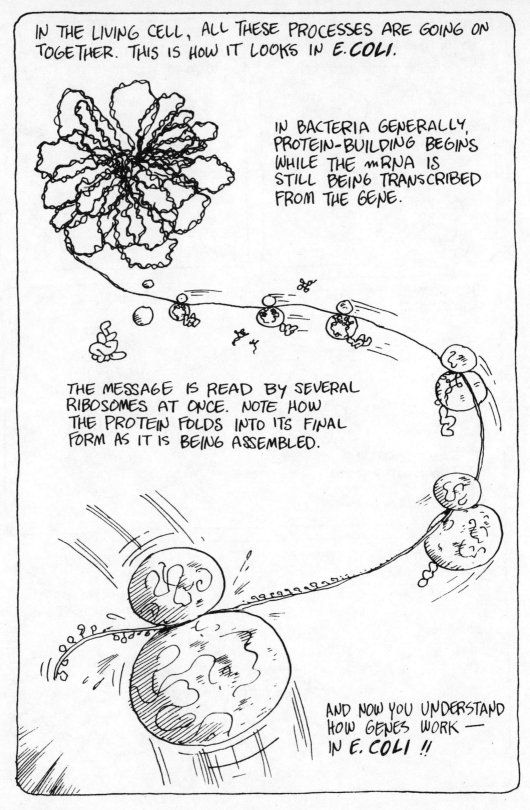

IN BACTERIA GENERALLY, PROTEIN-BUILDING BEGINS WHILE THE mRNA IS STILL BEING TRANSCRIBED FROM THE GENE.

THE MESSAGE IS READ BY SEVERAL RIBOSOMES AT ONCE. NOTE HOW THE PROTEIN FOLDS INTO ITS FINAL FORM AS IT IS BEING ASSEMBLED.

AND NOW YOU UNDERSTAND HOW GENES WORK — IN *E. COLI !!*

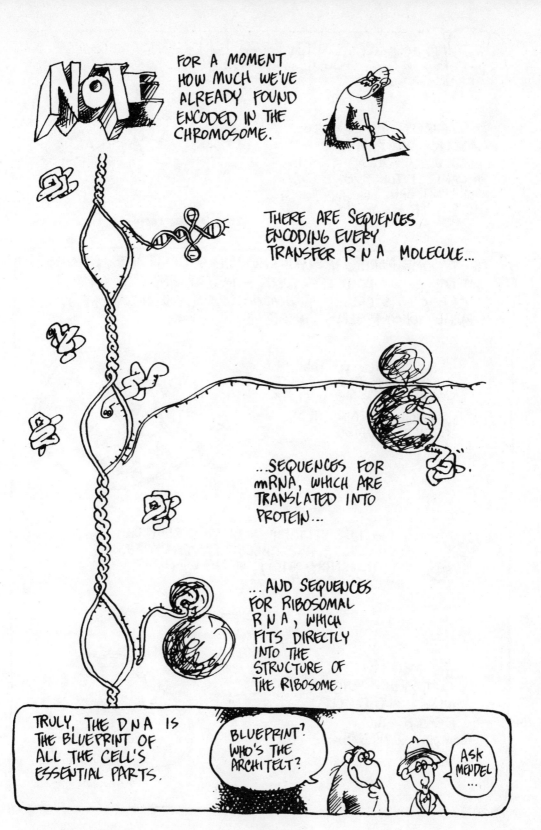

143

 # PRO AND EU

WE BEGAN BY ASKING ABOUT GORILLAS AND BANANAS, AND ENDED UP INSIDE SOME INSIGNIFICANT LITTLE BUG, E. COLI... NOW WHAT CAN WE SAY ABOUT OTHER LIFE FORMS?

THE HIGHER ONES?

FIRST, SOME MORE *JARGON*: THE CELLS OF PLANTS, ANIMALS, AND OTHER ADVANCED CREATURES — IN FACT, ANY CELL WITH A *NUCLEUS* — IS CALLED A *EUCARYOTE* ("YOU-CARRY-OAT"), MEANING "GOOD NUCLEUS" IN GREEK.

EUCARYOTES CONTAIN ALL SORTS OF BODIES, BUT THE KEY IS THE NUCLEUS, WHICH CONTAINS THE CHROMOSOMES.

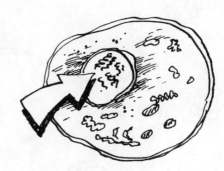

 THE TINY BACTERIA, WITH THEIR SIMPLER STRUCTURE, ARE CALLED *PROCARYOTES* ("PRO-CARRY-OATS"), MEANING "BEFORE NUCLEUS" IN GREEK.

THE IDEA IS THAT PROCARYOTES MUST HAVE EVOLVED *BEFORE* THE MORE COMPLICATED EUCARYOTES.

SONNY, IS THAT EU?

EUCARYOTES AND PROCARYOTES SHARE THE SAME BASIC GENETIC EQUIPMENT:

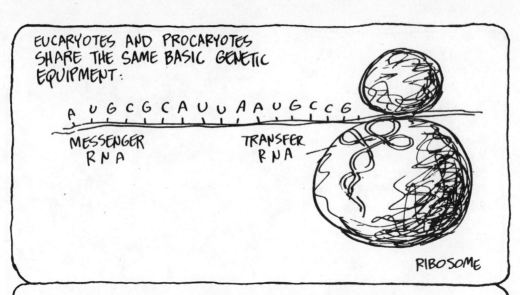

A U G C G C A U U A A U G C C G

MESSENGER RNA

TRANSFER RNA

RIBOSOME

AND ⟩ # IN ALL LIFE, THE GENETIC CODE IS THE SAME —

A FACT WHICH STRONGLY SUGGESTS THAT WE ALL COME FROM A COMMON ANCESTOR.

LET'S HAVE A FAMILY REUNION SOMETIME!

ANY TIME... I'LL BRING MY GORILLA...

BUT ▭ THERE ARE BIG DIFFERENCES BETWEEN PRO AND EU...

TO BEGIN WITH, EUCARYOTES HAVE ALL THEIR *RIBOSOMES* OUTSIDE THE NUCLEUS, SEPARATED FROM THE GENES BY A MEMBRANE.

HOW CAN YOU MAKE PROTEINS?

IT'S A BIT LIKE KISSING THROUGH PLASTIC...

145

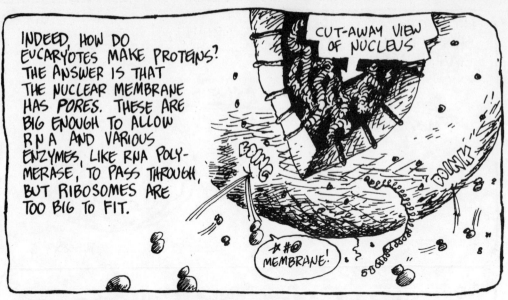

INDEED, HOW DO EUCARYOTES MAKE PROTEINS? THE ANSWER IS THAT THE NUCLEAR MEMBRANE HAS *PORES*. THESE ARE BIG ENOUGH TO ALLOW RNA AND VARIOUS ENZYMES, LIKE RNA POLYMERASE, TO PASS THROUGH, BUT RIBOSOMES ARE TOO BIG TO FIT.

CUT-AWAY VIEW OF NUCLEUS

*#@ MEMBRANE!

WITHIN THE NUCLEUS, mRNA IS MADE AS IN BACTERIA — BUT THEN COME CERTAIN MODIFICATIONS...

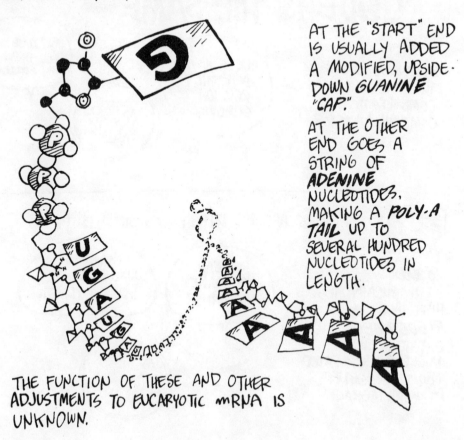

AT THE "START" END IS USUALLY ADDED A MODIFIED, UPSIDE-DOWN *GUANINE* "CAP."

AT THE OTHER END GOES A STRING OF *ADENINE* NUCLEOTIDES, MAKING A *POLY-A TAIL* UP TO SEVERAL HUNDRED NUCLEOTIDES IN LENGTH.

THE FUNCTION OF THESE AND OTHER ADJUSTMENTS TO EUCARYOTIC mRNA IS UNKNOWN.

THE NEXT MOVE CAME AS A GREAT SURPRISE TO GENETICISTS: A COMPLEX OF PROTEIN AND RNA GRABS THE mRNA, FORMING LOOPS, LIKE THIS —

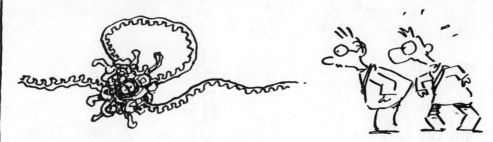

THE COMPLEX — CALLED A **SPLICEOSOME** — THEN SHEARS OFF THE LOOP, DISCARDS IT, SPLICES THE REMAINING PIECES TOGETHER, AND DEPARTS.

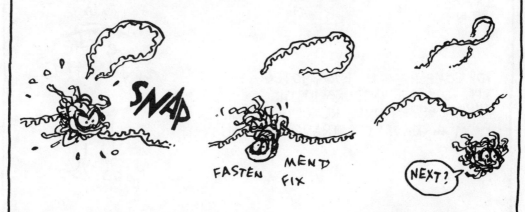

SNAP

FASTEN MEND FIX

NEXT?

THIS IS BIZARRE! EUCARYOTIC GENES CONTAIN "**JUNK DNA**" — NON-CODING MESSAGE SEQUENCES THAT HAVE TO BE CUT OUT BEFORE THE GENE CAN BE EXPRESSED!!

WHY?

TO KEEP SPLICEOSOMES BUSY?

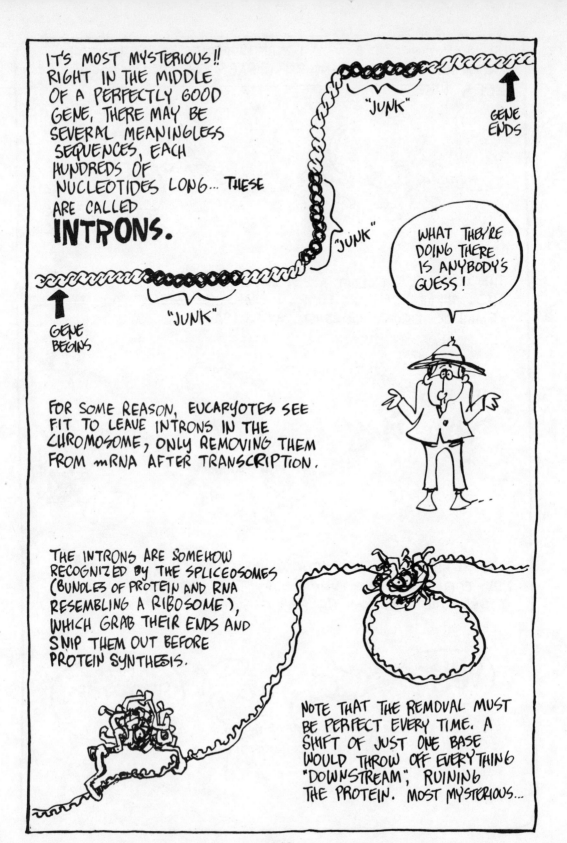

IT'S MOST MYSTERIOUS!! RIGHT IN THE MIDDLE OF A PERFECTLY GOOD GENE, THERE MAY BE SEVERAL MEANINGLESS SEQUENCES, EACH HUNDREDS OF NUCLEOTIDES LONG... THESE ARE CALLED **INTRONS.**

"JUNK"

GENE ENDS

"JUNK"

WHAT THEY'RE DOING THERE IS ANYBODY'S GUESS!

"JUNK"

GENE BEGINS

FOR SOME REASON, EUCARYOTES SEE FIT TO LEAVE INTRONS IN THE CHROMOSOME, ONLY REMOVING THEM FROM mRNA AFTER TRANSCRIPTION.

THE INTRONS ARE SOMEHOW RECOGNIZED BY THE SPLICEOSOMES (BUNDLES OF PROTEIN AND RNA RESEMBLING A RIBOSOME), WHICH GRAB THEIR ENDS AND SNIP THEM OUT BEFORE PROTEIN SYNTHESIS.

NOTE THAT THE REMOVAL MUST BE PERFECT EVERY TIME. A SHIFT OF JUST ONE BASE WOULD THROW OFF EVERYTHING "DOWNSTREAM", RUINING THE PROTEIN. MOST MYSTERIOUS...

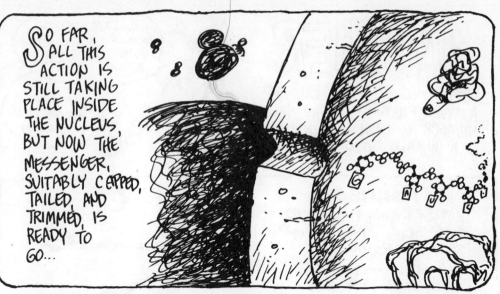

SO FAR, ALL THIS ACTION IS STILL TAKING PLACE INSIDE THE NUCLEUS, BUT NOW THE MESSENGER, SUITABLY CAPPED, TAILED, AND TRIMMED, IS READY TO GO...

AS IT PASSES THROUGH THE NUCLEAR MEMBRANE, THE RIBOSOMES BEGIN "READING OUT" THE PROTEIN, MUCH THE SAME AS IN PROCARYOTES.

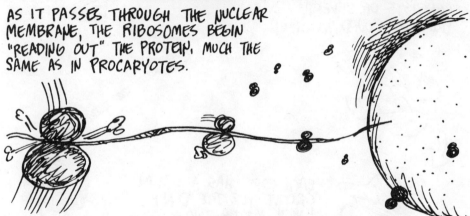

FINALLY THE PROTEIN GOES OFF TO DO ITS JOB; THE mRNA IS BROKEN DOWN INTO "SCRAP"; AND THE PARTS RETURN TO THE NUCLEUS FOR RECYCLING, TOGETHER WITH THE ENZYMES THAT DO THE JOB.

ANOTHER

DIFFERENCE BETWEEN EU AND A BACTERIUM IS IN THE SHEER NUMBER OF GENES: 200,000 IN A HUMAN, 4000 IN E. COLI.

HMM... 200,000 GENES... 1000 NUCLEOTIDES PER GENE... THAT'S 200 MILLION... MY MY!

HA! I HAVE THAT MANY SISTERS LIVING IN YOUR GUT!

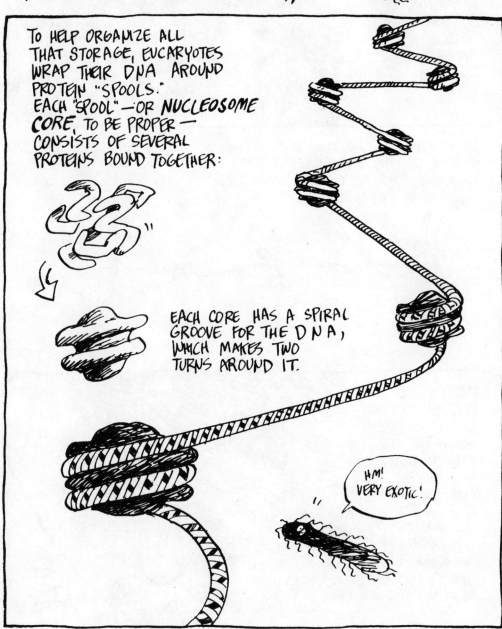

TO HELP ORGANIZE ALL THAT STORAGE, EUCARYOTES WRAP THEIR DNA AROUND PROTEIN "SPOOLS." EACH "SPOOL"—OR **NUCLEOSOME CORE**, TO BE PROPER—CONSISTS OF SEVERAL PROTEINS BOUND TOGETHER:

EACH CORE HAS A SPIRAL GROOVE FOR THE DNA, WHICH MAKES TWO TURNS AROUND IT.

HM! VERY EXOTIC!

WHEN A EUCARYOTIC CELL WANTS TO DIVIDE, DNA REPLICATION BEGINS AT MANY SITES AT ONCE (UNLIKE IN *E. COLI*, WHERE IT BEGINS AT ONE SITE).

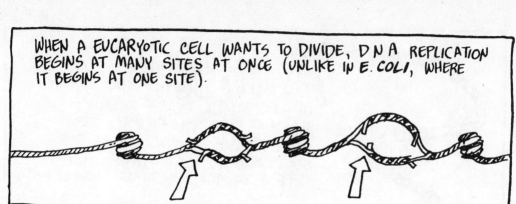

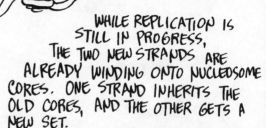

WHILE REPLICATION IS STILL IN PROGRESS, THE TWO NEW STRANDS ARE ALREADY WINDING ONTO NUCLEOSOME CORES. ONE STRAND INHERITS THE OLD CORES, AND THE OTHER GETS A NEW SET.

AS WE'VE SEEN, JUST BEFORE CELL DIVISION, THE CHROMOSOMES SHORTEN AND THICKEN. THIS MUST INVOLVE SOME WAY OF PACKING THE NUCLEOSOMES, BUT THE ARRANGEMENT AND PROCESS ARE STILL UNSOLVED PROBLEMS.

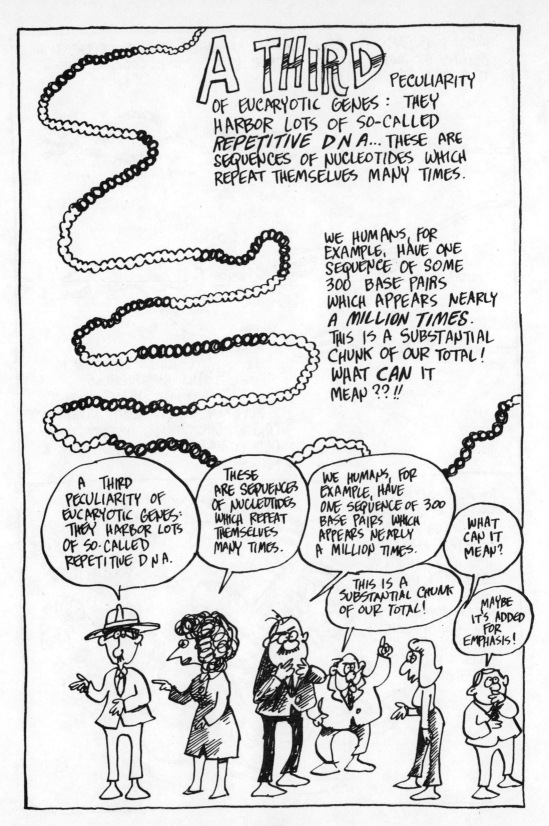

152

ONE POSSIBILITY IS THAT THEY COME FROM

VIRUSES

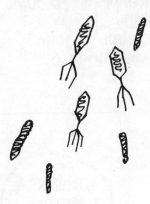

VIRUSES ARE THE SIMPLEST LIVING THINGS KNOWN— IF THEY'RE TRULY ALIVE AT ALL... THEY'RE SORT OF ALIVE AND NOT ALIVE...

REMINDS ME OF MY OLD BIOLOGY TEACHER...

EVEN SIMPLER AND SMALLER THAN A BACTERIUM, A VIRUS HAS ONLY TWO PARTS: A BIT OF *NUCLEIC ACID* WRAPPED UP IN A *PROTEIN COAT:*

CUT-AWAY VIEW

THE NUCLEIC ACID, WHICH MAY BE DNA OR RNA, ENCODES THE PROTEIN COAT AND A FEW ENZYMES NEEDED FOR REPLICATION.

BUT A VIRUS CAN'T REPRODUCE ON ITS OWN, BECAUSE IT LACKS RIBOSOMES AND THE REST OF A LIVING CELL'S PROTEIN-MAKING EQUIPMENT. A VIRUS CAN ONLY "LIVE" AS A **PARASITE**, BY INVADING A HOST CELL AND TAKING OVER ITS RIBOSOMES, ENZYMES, AND ENERGY.

VIRUSES LANDING ON A BACTERIUM, INJECTING IT WITH VIRAL DNA

ONCE IT GETS ITS DNA OR RNA INTO THE HOST, THE VIRUS BEGINS TO REPRODUCE WILDLY, STRAINING THE CELL TO THE BURSTING POINT!

THAT'S A TYPICAL LIFE-STYLE (OR NON-LIFE-STYLE) FOR A
VIRUS, BUT SOME VIRUSES ARE EVEN SNEAKIER: THEY
ACTUALLY INSERT THEIR GENES INTO THE HOST CELL'S DNA.

A **RETRO-VIRUS** IS AN
RNA VIRUS ENCODING
AN ENZYME THAT
MAKES A DNA COPY
OF ITS RNA AND
SPLICES IT INTO THE
HOST CHROMOSOME.

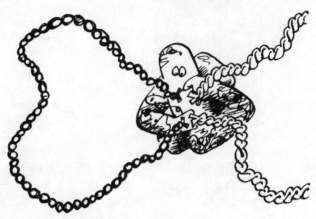

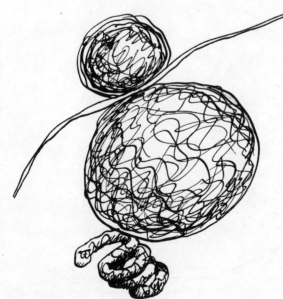

THIS IS ONE REASON
WHY SOME VIRAL
INFECTIONS ARE
INCURABLE: THE VIRUS'
GENES CAN'T BE
GOTTEN RID OF.
YOUR OWN CHROMOSOMES
MAY BE DIRECTING
THE PRODUCTION OF
MORE VIRUSES!!! THE
AIDS VIRUS WORKS
THIS WAY.

IT'S POSSIBLE THAT SOME OF THE REPETITIVE AND "JUNK" D N A IN OUR CHROMOSOMES MAY HAVE COME FROM THIS SOURCE: ANCIENT VIRUSES THAT MANAGED TO INSERT THEIR HEREDITARY BLUEPRINT INTO OUR ANCESTORS' D N A.

SUBVERSIVE ELEMENTS!

IF SO, THE "EDITING" OF mRNA MAY HAVE EVOLVED AS A DEFENSE AGAINST INAPPROPRIATE SEQUENCES STUCK INTO THE MIDDLE OF GENES.

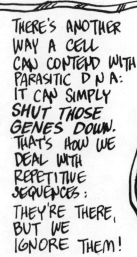

THERE'S ANOTHER WAY A CELL CAN CONTEND WITH PARASITIC D N A: IT CAN SIMPLY *SHUT THOSE GENES DOWN.* THAT'S HOW WE DEAL WITH REPETITIVE SEQUENCES: THEY'RE THERE, BUT WE IGNORE THEM!

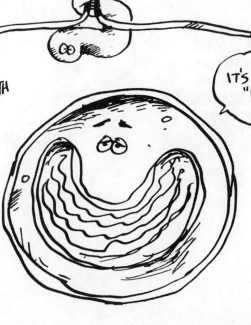

IT'S CALLED "REPRESSIVE TOLERANCE."

THE BATTLE AGAINST VIRUSES IS NEVER-ENDING...

Mutation & Dominance

(again!)

NOW THAT WE KNOW WHAT GENES REALLY ARE, WE CAN GET A MUCH BETTER GRASP OF *MUTATION* AND *DOMINANCE*.

A MUTATION IN A GENE IS JUST A CHANGE IN THE DNA'S SEQUENCE OF NUCLEOTIDES. EVEN A MISTAKE AT JUST *ONE POSITION* CAN HAVE A PROFOUND EFFECT.

HERE IS A SMALL BUT DEVASTATING MUTATION IN THE GENE FOR *HEMOGLOBIN*, THE PROTEIN WHICH CARRIES OXYGEN IN THE BLOOD.

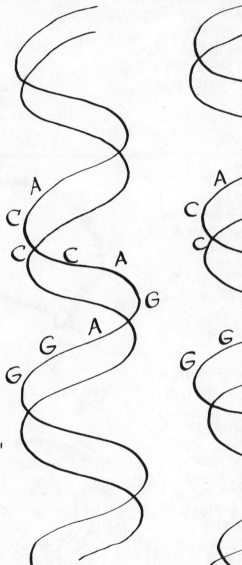

GOOD GENE   MUTANT GENE

THE REASON, OF COURSE, IS THAT THE CHANGE IS REFLECTED IN THE *PROTEIN* WHICH THE GENE ENCODES... FIRST THE mRNA COMES OUT WRONG, AND THEN THE PROTEIN...

RIGHT

A A C C A G

⇩

GLN

WRONG

A A C U A G

⇩

"STOP"

THIS ESPECIALLY DISASTROUS MUTATION, WHICH INTERRUPTS THE PROTEIN IN THE MIDDLE, CAUSES A SERIOUS CONDITION CALLED *THALASSEMIA*, AN INABILITY TO MAKE HEMOGLOBIN. THE VICTIM SUFFERS FROM A PAINFUL LACK OF OXYGEN.

SOMETIMES A CHANGE MAY MAKE NO DIFFERENCE AT ALL. IF YOU REFER BACK TO THE CODE TABLE, YOU'LL RECALL THAT IT'S SOMEWHAT *REDUNDANT* — MEANING THAT ONE AMINO ACID MAY BE ENCODED BY SEVERAL DIFFERENT CODONS.

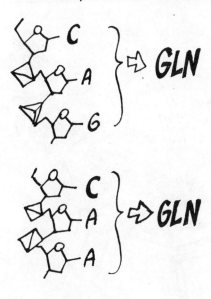

C A G ⇨ **GLN**

C A A ⇨ **GLN**

THIS IS A KIND OF DEFENSE AGAINST MUTATIONS!

OCCASIONALLY, THE "MISTAKEN" AMINO ACID MAY FIT IN FAIRLY WELL (THOUGH USUALLY LESS THAN PERFECTLY).

ODD... I SEEM TO HAVE LOST SOME *BITE*...

SOMETIMES — ONCE IN A BLUE MOON — THE PROTEIN MAY EVEN WORK *BETTER* THAN BEFORE.

GOSH!

BUT MOST OF THE TIME, A MUTATION JUST RUINS THE PROTEIN. IT'S MUCH EASIER TO MESS SOMETHING UP THAN TO IMPROVE IT! IF YOU DOUBT IT, TRY MAKING RANDOM CHANGES IN SOME HOUSEHOLD APPLIANCE!!

THERE GOES THE — *UNGH* — COMPUTER!

 **EARLIER** (p. 81)

WE NOTED THAT MOST MUTATIONS ARE *RECESSIVE*. NOW WE CAN SEE WHY: A MUTATION USUALLY CAUSES AN *INABILITY* TO MAKE AN ENZYME. IN THE EXAMPLE ABOVE, THE MUTANT GENE FAILED TO MAKE HEMOGLOBIN.

HOWEVER, WE HAVE *TWO SETS* OF CHROMOSOMES. EVEN IF A MUTATION AFFECTS ONE OF THEM, THE "INSURANCE" GENE WILL STILL PRODUCE ITS ENZYME.

GOOD GENE
⇓
HEMOGLOBIN

BAD GENE
⇓
NO HEMOGLOBIN

ONLY THE UNLUCKY SOUL WITH A DOUBLE DOSE OF MUTANT GENES WILL BE AFFLICTED WITH THALASSEMIA.

WHEN YOUR GENETIC INSURANCE LAPSES, THAT'S WHEN YOU'D BETTER GET MEDICAL INSURANCE!

WE DIDN'T MENTION IT EARLIER, BUT SOME ALLELES CAN BE

CO-DOM·IN·ANT,

MEANING THAT A HETEROZYGOTE MAKES BOTH PHENOTYPES. AN EXAMPLE IS *BLOOD GROUPS.*

AH, I LOVE VARIETY...

THERE IS A GENETICALLY DETERMINED SEQUENCE OF SUGARS LYING ON THE SURFACE OF RED BLOOD CELLS. ONE ALLELE, I^A, MAKES SEQUENCE A. ANOTHER ALLELE, I^B, MAKES SEQUENCE B.

A   B

$I^A I^A$

IF HOMOZYGOUS FOR I^A YOUR BLOOD HAS ONLY SEQUENCE A. THIS IS *TYPE A BLOOD.*

$I^B I^B$

IF HOMOZYGOUS FOR I^B, YOU HAVE *TYPE B BLOOD.*

$I^A I^B$

A HETERO-ZYGOTE MAKES BOTH SEQUENCES, AND HAS *TYPE AB BLOOD.*

AND FINALLY, THERE IS A THIRD ALLELE, I^O, MAKING NO SUGAR SEQUENCE. TYPE O BLOOD IS *RECESSIVE.*

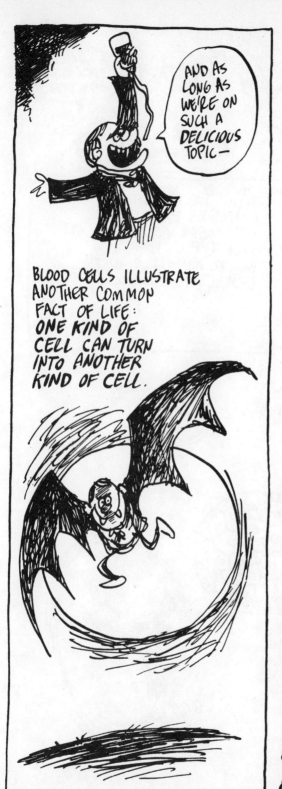

AND AS LONG AS WE'RE ON SUCH A DELICIOUS TOPIC—

BLOOD CELLS ILLUSTRATE ANOTHER COMMON FACT OF LIFE: ONE KIND OF CELL CAN TURN INTO ANOTHER KIND OF CELL.

A RED BLOOD CELL BEGINS ITS EXISTENCE AS A BONE MARROW CELL, A PERFECTLY GOOD EUCARYOTE, BUT LACKING IN HEMOGLOBIN.

AT SOME POINT, A MARROW CELL BEGINS TO CHANGE... AMONG OTHER THINGS, IT BEGINS TO MAKE HEMOGLOBIN.

EVENTUALLY, IT EMERGES AS A FULLY DEVELOPED RED BLOOD CELL.

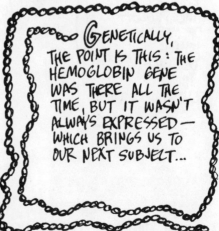

GENETICALLY, THE POINT IS THIS: THE HEMOGLOBIN GENE WAS THERE ALL THE TIME, BUT IT WASN'T ALWAYS EXPRESSED— WHICH BRINGS US TO OUR NEXT SUBJECT...

GENE REGULATION

> SORRY— YOU CAN'T PARK THAT GENE HERE—

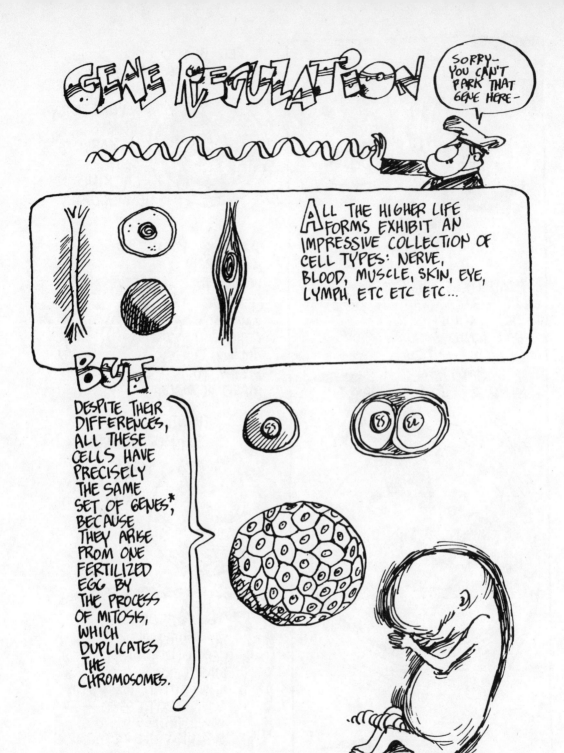

All the higher life forms exhibit an impressive collection of cell types: nerve, blood, muscle, skin, eye, lymph, etc etc etc...

BUT

despite their differences, all these cells have precisely the same set of genes,* because they arise from one fertilized egg by the process of mitosis, which duplicates the chromosomes.

*As usual, there are exceptions!!

CLEARLY, DIFFERENT GENES COME INTO PLAY IN DIFFERENT CELLS... SO EACH CELL MUST HAVE WAYS OF "DECIDING" WHICH GENES TO "TURN ON" AND WHEN TO DO IT...

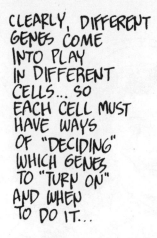

OTHERWISE, ONE DREADS THE RESULTS!

EVEN THE LOWLY BACTERIUM NEEDS TO REGULATE ITS GENES. WHEN FOOD IS AVAILABLE, IT NEEDS TO MAKE ENZYMES TO DIGEST IT; WHEN IT RUNS LOW ON AN AMINO ACID, IT HAS TO SYNTHESIZE MORE; ETC ETC ETC...

AS USUAL, THE QUESTION HAS BEEN MOST THOROUGHLY STUDIED IN *E. COLI.*

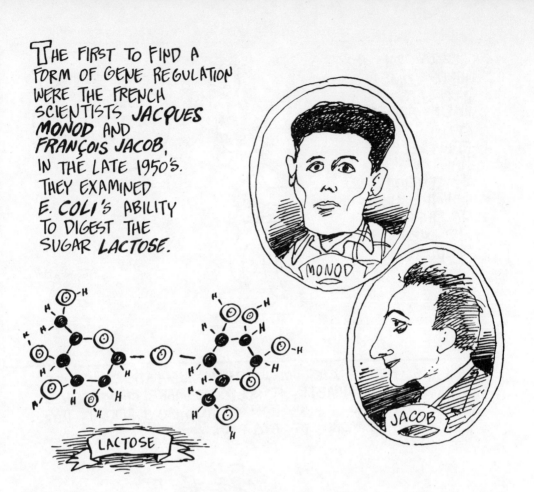

THE FIRST TO FIND A FORM OF GENE REGULATION WERE THE FRENCH SCIENTISTS **JACQUES MONOD** AND **FRANÇOIS JACOB**, IN THE LATE 1950's. THEY EXAMINED _E. COLI_'s ABILITY TO DIGEST THE SUGAR **LACTOSE**.

LACTOSE

MONOD

JACOB

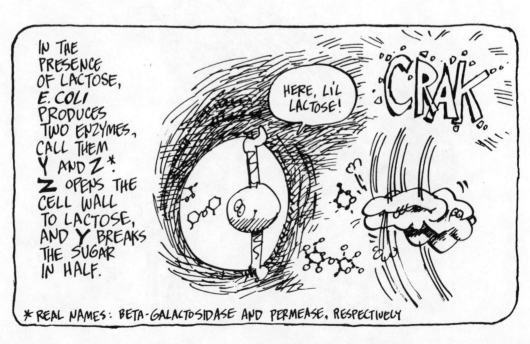

IN THE PRESENCE OF LACTOSE, E. COLI PRODUCES TWO ENZYMES, CALL THEM Y AND Z* Z OPENS THE CELL WALL TO LACTOSE, AND Y BREAKS THE SUGAR IN HALF.

HERE, LI'L LACTOSE!

CRAK

* REAL NAMES: BETA-GALACTOSIDASE AND PERMEASE, RESPECTIVELY

WITHOUT GOING INTO THE DETAILS OF THEIR EXPERIMENTS, WHICH WERE QUITE INVOLVED, HERE ARE SOME OF MONOD AND JACOB'S MAIN RESULTS:

THIS EXPERIMENT WAS MORE DIFFICULT THAN A CHEESE SOUFFLÉ!

FIRST, THEY FOUND THAT THE GENES FOR Y AND Z, CALLED "lac Y" AND "lac Z," LAY TOGETHER, SIDE-BY-SIDE, ON THE CHROMOSOME. SUCH A CLUSTER OF GENES, ENCODING RELATED ENZYMES, AND REGULATED TOGETHER, IS CALLED AN

OPERON:

THIS IS THE "lac OPERON":

lac Y

AH LAC TH' GRAND OLE OPERON!

← lac P →|← lac O →|← ..... lac Z

WE'RE ABOUT TO EXPLAIN THIS PART!

AT THE START OF THIS (AND EVERY) OPERON IS A *PROMOTER* REGION, HERE CALLED lac P. THIS IS THE SITE WHERE THE ENZYME *RNA POLYMERASE* BINDS ONTO THE DNA TO BEGIN TRANSCRIBING THE MESSAGE INTO mRNA. (SEE p. 133.)

MMM!

167

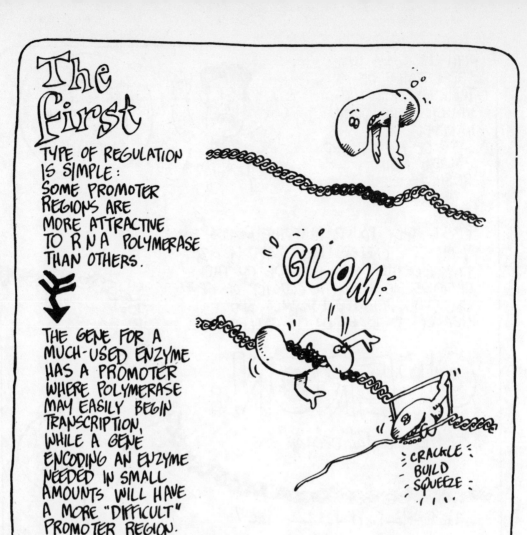

The First

TYPE OF REGULATION IS SIMPLE: SOME PROMOTER REGIONS ARE MORE ATTRACTIVE TO RNA POLYMERASE THAN OTHERS.

THE GENE FOR A MUCH-USED ENZYME HAS A PROMOTER WHERE POLYMERASE MAY EASILY BEGIN TRANSCRIPTION, WHILE A GENE ENCODING AN ENZYME NEEDED IN SMALL AMOUNTS WILL HAVE A MORE "DIFFICULT" PROMOTER REGION.

GLOM

CRACKLE
BUILD
SQUEEZE

WHAT ABOUT THE LACTOSE OPERON, WHOSE ENZYMES ARE SOMETIMES NEEDED IN QUANTITY (WHEN LACTOSE IS PRESENT), BUT OTHERWISE NOT NEEDED AT ALL ??

lac O

THAT'S WHERE lac O COMES IN!

O?

MONOD & JACOB'S IDEA:
THERE IS A PROTEIN,
THE

REPRESSOR,

WHICH SITS ON THE DNA
AT A SPOT BETWEEN
THE PROMOTER AND
THE FIRST GENE, lac Z.
THIS SPOT IS CALLED
THE OPERATOR,
lac O.

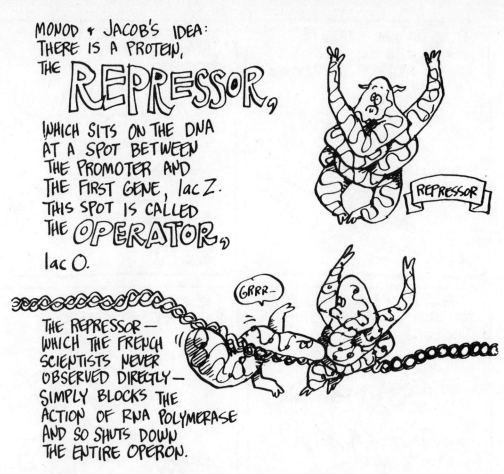

REPRESSOR

GRRR—

THE REPRESSOR—
WHICH THE FRENCH
SCIENTISTS NEVER
OBSERVED DIRECTLY—
SIMPLY BLOCKS THE
ACTION OF RNA POLYMERASE
AND SO SHUTS DOWN
THE ENTIRE OPERON.

ONE MORE THING ABOUT THE REPRESSOR: IT CAN ALSO BIND
TO **LACTOSE***— BUT DOING SO CAUSES THE REPRESSOR TO
"FLEX" AND RELEASE THE DNA:

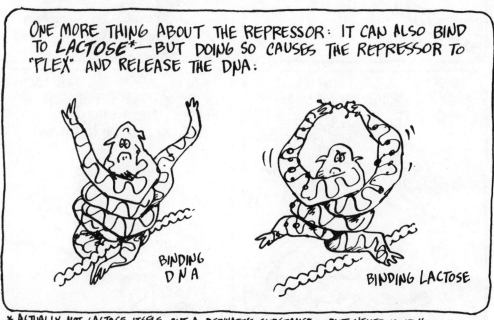

BINDING
DNA

BINDING LACTOSE

* ACTUALLY NOT LACTOSE ITSELF, BUT A DERIVATIVE SUBSTANCE — BUT NEVER MIND !!

IN THE NORMAL STATE OF AFFAIRS, THE REPRESSOR SITS ON THE OPERATOR, REPRESSING THE GENE:

ALONG COMES A LITTLE LACTOSE, ATTRACTING THE REPRESSOR:

IT FLEXES, GRASPING THE SUGAR, AND RNA POLYMERASE SLIPS THROUGH!!

THE ENTIRE OPERON IS THEN EXPRESSED REPEATEDLY.

THE NEWLY MADE PROTEINS BRING IN MORE LACTOSE AND DIGEST IT...

FINALLY, WHEN ALL THE LACTOSE IS GONE, THE REPRESSOR UN-FLEXES AND RETURNS TO ITS SPOT ON THE CHROMOSOME.

REPRESSORS TURN OUT TO BE A COMMON WAY TO REGULATE "INDUCIBLE" ENZYMES—I.E., ENZYMES WHICH ARE MADE IN RESPONSE TO A CHEMICAL-LIKE LACTOSE... BUT DESPITE THIS BRILLIANT IDEA, MONOD AND JACOB COULD NEVER ACTUALLY FIND A REPRESSOR. IT REMAINED A THEORETICAL POSSIBILITY...

THESE REPRESSORS ARE MORE ELUSIVE THAN A PERFECT SAUCE BÉARNAISE...

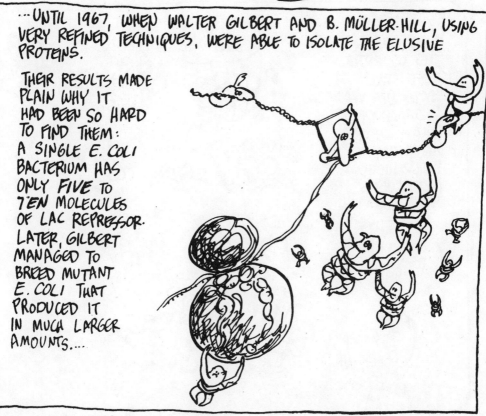

...UNTIL 1967, WHEN WALTER GILBERT AND B. MÜLLER-HILL, USING VERY REFINED TECHNIQUES, WERE ABLE TO ISOLATE THE ELUSIVE PROTEINS.

THEIR RESULTS MADE PLAIN WHY IT HAD BEEN SO HARD TO FIND THEM: A SINGLE E. COLI BACTERIUM HAS ONLY FIVE TO TEN MOLECULES OF LAC REPRESSOR. LATER, GILBERT MANAGED TO BREED MUTANT E. COLI THAT PRODUCED IT IN MUCH LARGER AMOUNTS....

ATTENUATION

ANOTHER METHOD OF GENE REGULATION GOES BY THE NAME OF:

AND ITS SUCCESSOR, ELEVEN-TUATION!

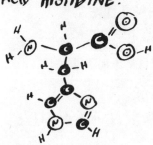

THIS GOVERNS AN E. COLI OPERON RESPONSIBLE FOR CONSTRUCTING THE AMINO ACID *HISTIDINE*.

WHEN E. COLI RUNS LOW ON THIS ESSENTIAL STUFF, THE BACTERIUM PRODUCES A GROUP OF *NINE* PROTEINS, WHICH CAN BUILD HISTIDINE MOLECULES FROM SCRATCH.

..AUTOMATIC ENZYMATIC...

AN ENZYMATIC ASSEMBLY LINE!

AS BEFORE, ALL 9 ENZYMES HAVE THEIR GENES CLUSTERED INTO AN OPERON, WITH AN INITIAL PROMOTER REGION. UNLIKE THE LAC OPERON, THIS ONE HAS NO PLACE FOR A REPRESSOR.

O GOOD!

INSTEAD, THERE IS A "LEADER SEQUENCE" ENCODING A PEPTIDE RICH IN HISTIDINE— THE VERY STUFF WE'RE TRYING TO MANUFACTURE.

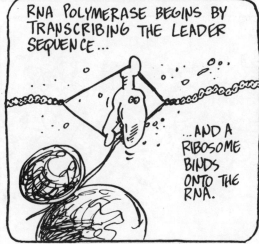

RNA POLYMERASE BEGINS BY TRANSCRIBING THE LEADER SEQUENCE...

...AND A RIBOSOME BINDS ONTO THE RNA.

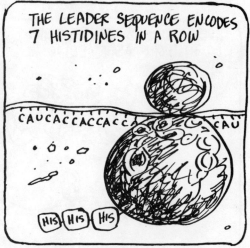

THE LEADER SEQUENCE ENCODES 7 HISTIDINES IN A ROW

CAUCACCACCACC ... CAU

HIS HIS HIS

IF HISTIDINE IS PLENTIFUL, THE RIBOSOME ZIPS ALONG, AND A LOOP FORMS IN THE mRNA.

BUMP

THIS LOOP BUMPS THE RNA POLYMERASE OFF THE OPERON, HALTING TRANSCRIPTION.

IF, ON THE OTHER HAND, HISTIDINE IS IN SHORT SUPPLY, THE RIBOSOME FALLS BEHIND THE POLYMERASE.

IN THIS CASE, A **DIFFERENT** LOOP FORMS, WHICH, BY PREVENTING THE FIRST LOOP, ENABLES THE POLYMERASE TO GO ON, AND THE OPERON IS EXPRESSED!

THE NEWLY MADE PROTEINS GO TO WORK ASSEMBLING HISTIDINE.

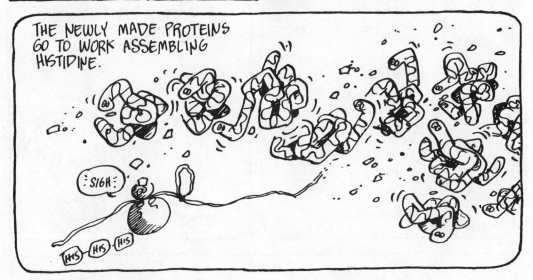

SIGH

HIS HIS HIS

RESULT?

A SHORTAGE OF HISTIDINE TURNS THE GENE **ON**, WHILE A HISTIDINE GLUT TURNS IT **OFF**.

CHEW ON THAT!

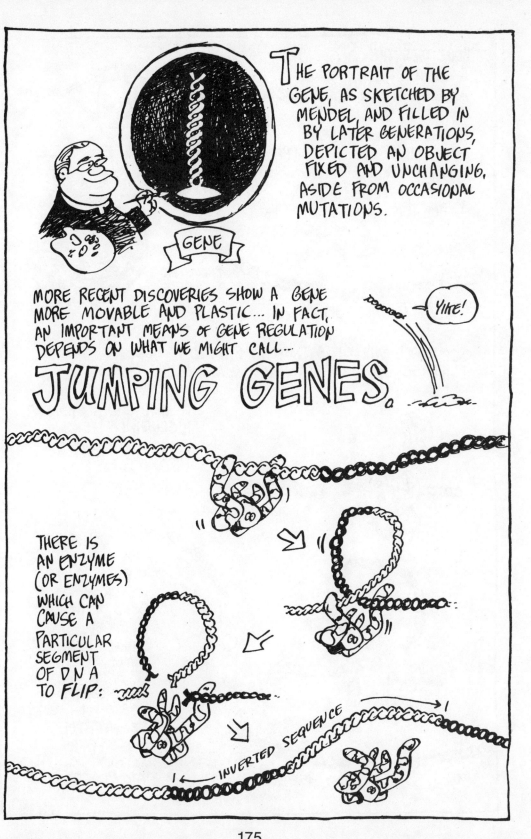

THE PORTRAIT OF THE GENE, AS SKETCHED BY MENDEL, AND FILLED IN BY LATER GENERATIONS, DEPICTED AN OBJECT FIXED AND UNCHANGING, ASIDE FROM OCCASIONAL MUTATIONS.

GENE

MORE RECENT DISCOVERIES SHOW A GENE MORE MOVABLE AND PLASTIC... IN FACT, AN IMPORTANT MEANS OF GENE REGULATION DEPENDS ON WHAT WE MIGHT CALL...

YIKE!

JUMPING GENES.

THERE IS AN ENZYME (OR ENZYMES) WHICH CAN CAUSE A PARTICULAR SEGMENT OF DNA TO FLIP:

INVERTED SEQUENCE

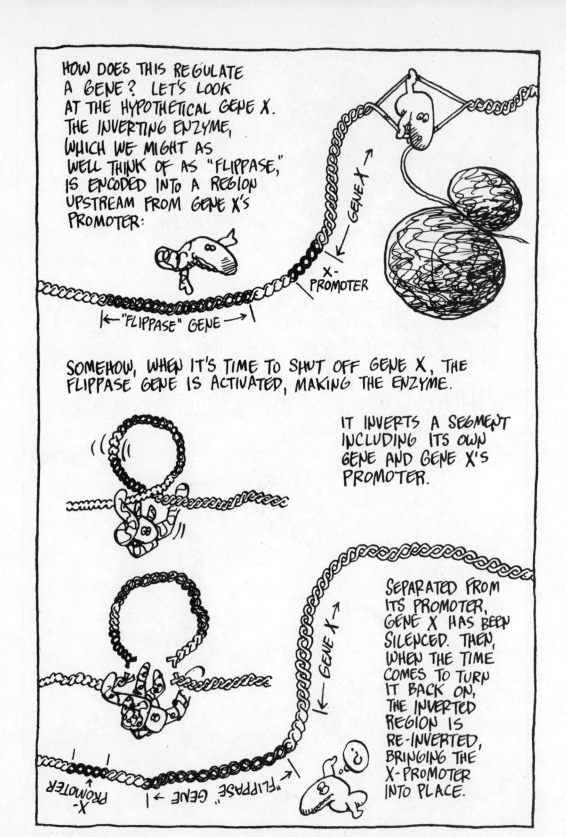

HOW DOES THIS REGULATE A GENE? LET'S LOOK AT THE HYPOTHETICAL GENE X. THE INVERTING ENZYME, WHICH WE MIGHT AS WELL THINK OF AS "FLIPPASE," IS ENCODED INTO A REGION UPSTREAM FROM GENE X'S PROMOTER:

|← "FLIPPASE" GENE →|

X- PROMOTER

GENE X →

SOMEHOW, WHEN IT'S TIME TO SHUT OFF GENE X, THE FLIPPASE GENE IS ACTIVATED, MAKING THE ENZYME.

IT INVERTS A SEGMENT INCLUDING ITS OWN GENE AND GENE X'S PROMOTER.

SEPARATED FROM ITS PROMOTER, GENE X HAS BEEN SILENCED. THEN, WHEN THE TIME COMES TO TURN IT BACK ON, THE INVERTED REGION IS RE-INVERTED, BRINGING THE X-PROMOTER INTO PLACE.

←GENE X →

|← "FLIPPASE" GENE →|

X- PROMOTER

SUCH MOVABLE
SECTIONS, OR
TRANSPOSONS,
ARE COMMON IN BOTH
PROCARYOTES AND
EUCARYOTES. BESIDES
INVERTING, THEY CAN
JUMP FROM PLACE
TO PLACE, FROM
CHROMOSOME TO
CHROMOSOME. THE
FULL FUNCTION OF
TRANSPOSONS IS
STILL A MYSTERY.

THE MOST SPECTACULAR EXAMPLES OF JUMPING GENES ARE
THE ONES ENCODING *ANTIBODIES.*

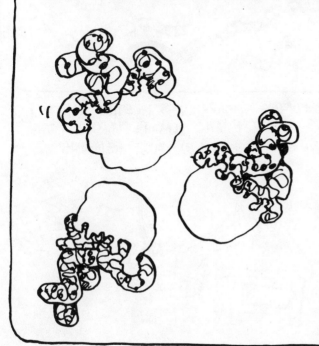

ANTIBODIES ARE
PROTEINS WHICH
SERVE AS THE BODY'S
DEFENSIVE WEAPONS.
THEY ATTACK
BACTERIA, VIRUSES,
AND OTHER
HARMFUL INVADERS.
THERE ARE LITERALLY
BILLIONS OF
POTENTIAL ANTIBODIES,
EACH KEYED TO
THE EXACT SHAPE
OF ITS "ENEMY."
HOW CAN SO MANY
BE ENCODED IN
GENES?

RATHER THAN HAVING BILLIONS OF GENES FOR ANTIBODIES, THE CHROMOSOMES CARRY A "TOOL KIT" OF A FEW HUNDRED *PARTIAL GENES.*

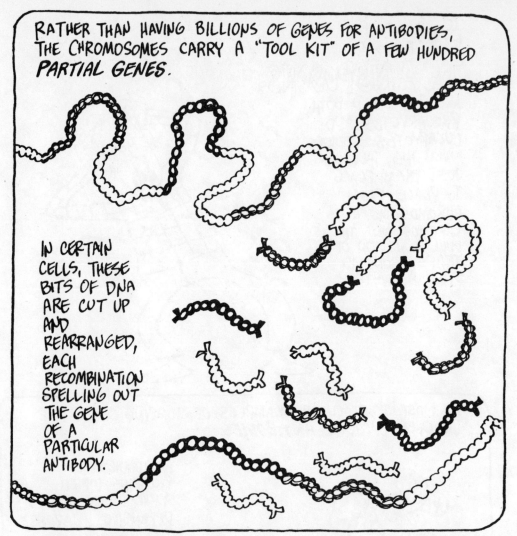

IN CERTAIN CELLS, THESE BITS OF DNA ARE CUT UP AND REARRANGED, EACH RECOMBINATION SPELLING OUT THE GENE OF A PARTICULAR ANTIBODY.

HOW THE ORGANISM REGULATES THIS PROCESS IS STILL A RIDDLE, AS ARE MOST MATTERS OF EUCARYOTIC GENE REGULATION: THE QUESTION OF HEMOGLOBIN (p. 163), FOR EXAMPLE, REMAINS WITHOUT AN ANSWER.

IT'S CLEAR THAT THE FLEXIBLE GENES OF EUCARYOTES WILL BE AN ACTIVE AREA OF RESEARCH IN YEARS TO COME.

GENETIC ENGINEERING

FOR ONE THING, PEOPLE CAN NOW *SPLICE* TWO PIECES OF DNA IN THE TEST TUBE — JUST LIKE SPLICING FILM...

HMM... I'LL ATTACH "GOD'S LITTLE ACRE" TO "VIVA VILLA."...

I'LL CALL IT "GODZILLA!"

THE COMBINATIONS CAN BE PRETTY BIZARRE: MOST COMMONLY, HUMAN GENES ARE ATTACHED TO THOSE OF A BACTERIUM, LIKE E. COLI...

WHAT ARE YOU — A MAN OR A MICROBE?

THIS IS WHAT YOU CALL

RECOMBINANT DNA

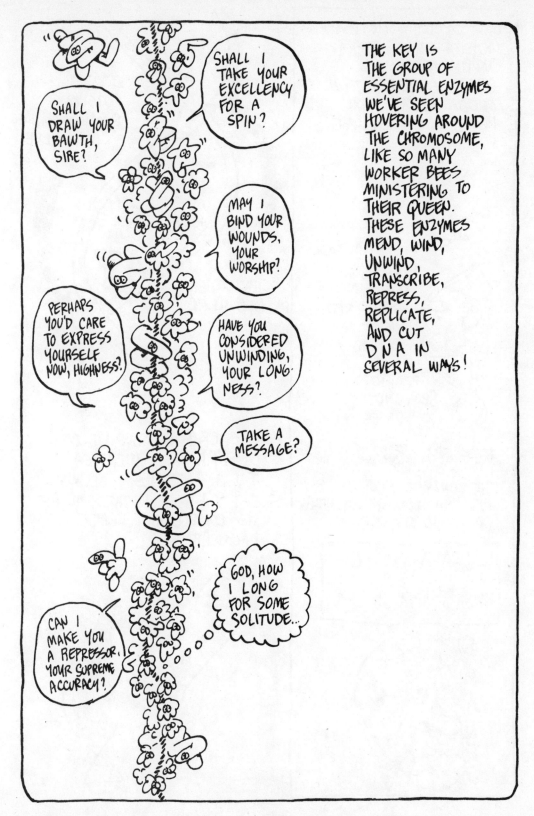

THE KEY IS THE GROUP OF ESSENTIAL ENZYMES WE'VE SEEN HOVERING AROUND THE CHROMOSOME, LIKE SO MANY WORKER BEES MINISTERING TO THEIR QUEEN. THESE ENZYMES MEND, WIND, UNWIND, TRANSCRIBE, REPRESS, REPLICATE, AND CUT D N A IN SEVERAL WAYS!

GENE SPLICING DEPENDS ON A SPECIAL TYPE OF CUTTING ENZYME CALLED A **RESTRICTION ENDONUCLEASE**, OR RESTRICTION ENZYME FOR SHORT.

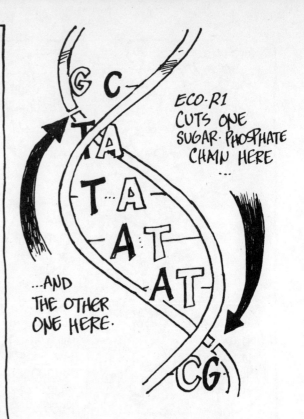

ECO-R1 CUTS ONE SUGAR-PHOSPHATE CHAIN HERE...

...AND THE OTHER ONE HERE.

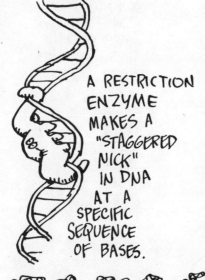

A RESTRICTION ENZYME MAKES A "STAGGERED NICK" IN DNA AT A SPECIFIC SEQUENCE OF BASES.

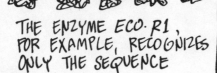

THE ENZYME ECO-R1, FOR EXAMPLE, RECOGNIZES ONLY THE SEQUENCE

-G-A-A-T-T-C-
-C-T-T-A-A-G-

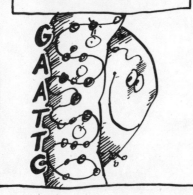

THIS CREATES TWO PIECES OF DNA WITH IDENTICAL T-T-A-A "TAILS." (BECAUSE C-T-T-A-A-G IS THE SAME AS ITS COMPLEMENT READ BACKWARDS!)

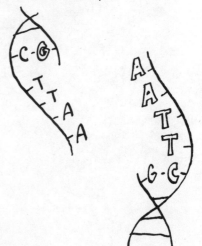

E. COLI USES ECO·R1 TO CHOP UP "ENEMY" VIRAL DNA, BUT HUMANS HAVE PUT IT TO CONSTRUCTIVE USE.

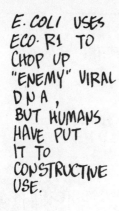

YOU'RE BEATING MY SWORDS INTO PLOWSHARES?

NO... *STOCK SHARES* IN GENETIC ENGINEERING COMPANIES..

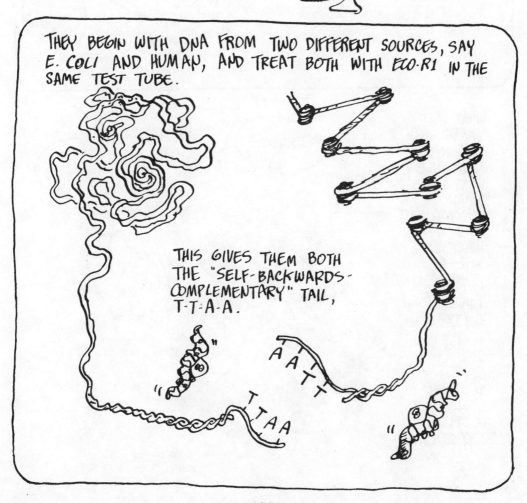

THEY BEGIN WITH DNA FROM TWO DIFFERENT SOURCES, SAY E. COLI AND HUMAN, AND TREAT BOTH WITH ECO·R1 IN THE SAME TEST TUBE.

THIS GIVES THEM BOTH THE "SELF-BACKWARDS-COMPLEMENTARY" TAIL, T·T·A·A.

A A T T

T T A A

THE TAILS SNAP TOGETHER, AND, AFTER TREATMENT WITH *LIGASE*, AN ENZYME THAT SEALS NICKS IN THE SUGAR-PHOSPHATE CHAIN, THE *RECOMBINANT DNA* IS COMPLETE!

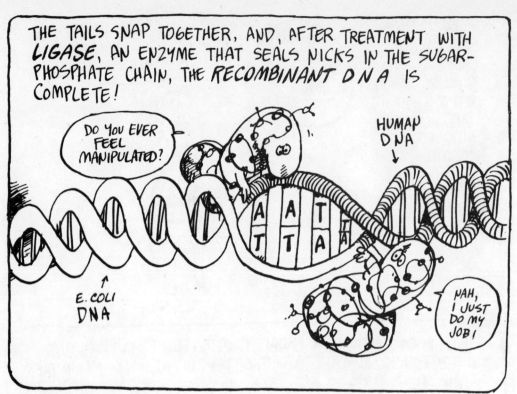

WHAT CAN YOU DO WITH THIS HYBRID MOLECULE? WHAT HAPPENS WHEN RECOMBINANT DNA IS INSERTED INTO A LIVING SYSTEM?

UNDER SOME CONDITIONS, IT TURNS OUT THAT GENE SPLICING CAN BE USEFUL IN *PRACTICE...*

THE TECHNIQUE IS CALLED

GENE CLONING,

AND IT WORKS LIKE THIS:

FIRST, CHOOSE A HUMAN GENE ENCODING SOME USEFUL PROTEIN.

IS THERE A PROTEIN THAT PUTS YOU THROUGH MEDICAL SCHOOL?

FOR YOUR BACTERIAL DNA, YOU NEED SOMETHING THAT WILL BE REPLICATED ONCE IT'S RETURNED TO THE CELL — A "*VECTOR*", SO-CALLED.

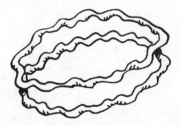

LUCKILY, *E. COLI* HAS SMALL RINGS OF DNA CALLED *PLASMIDS*, SEPARATE FROM THE CHROMOSOME. YOU CHOOSE (OR ENGINEER!) A PLASMID CONTAINING THE SEQUENCE G·A·A·T·T·C, AND REMOVE IT FROM THE BACTERIUM.

JUST AS ABOVE, YOU *SPLICE* THE HUMAN GENE INTO THE PLASMID —

AND PUT IT BACK INTO *E. COLI*.

MAHSTER! MAHSTER! ALIEN SEQUENCES! WHAT DO WE DO?

EXPRESS IT, AND SEE WHAT IT WANTS!

BE FRUITFUL AND DIVIDE!

NOW YOU FEED THE BACTERIUM AND LET IT BREED.

THE PLASMID IS REPLICATED ALONG WITH EVERYTHING ELSE IN THE BACTERIUM.

WITHIN A FEW HOURS, WE CAN HAVE A **BILLION** BACTERIA IN A FEW DROPS OF CULTURE— AND A BILLION COPIES OF THE HUMAN GENE !!

IF WE'VE INCLUDED THE PROPER REGULATORY REGIONS AS WELL, THE BACTERIA SHOULD EXPRESS THE GENE, AND WE CAN EXTRACT SUBSTANTIAL AMOUNTS OF THE HUMAN PROTEIN. MIRACULOUS!

PRAISE BE!

INTERFER

INSULIN

THE PROCEDURE SOUNDS SIMPLE — AND, IN PRINCIPLE, IT IS. IN PRACTICE IT CAN BE MOST COMPLICATED, BUT THE FOLKS IN THE LABS HAVE SOLVED MOST OF THOSE PRACTICAL PROBLEMS. WE CAN NOW CLONE JUST ABOUT ANY GENE WE WANT... USUALLY IN *E. COLI*, BUT OTHER FAST-GROWING ORGANISMS WORK, AS WELL, EVEN EUCARYOTES LIKE *YEAST* —

HI! NOT ONLY DO I FULFILL THE MINIMUM DAILY REQUIREMENT OF VITAMINS A, B, C, D, AND K, BUT ALSO I TASTE LIKE ROAST DUCKLING AND PREVENT CANCER!

'BREAD OF THE FUTURE!'

IT'S EVEN POSSIBLE TO CLONE GENES INTO HUMAN CELLS, BUT SO FAR IT ONLY WORKS IN A DISH, NOT IN A REAL PERSON...

BUT ONE OF THESE DAYS...

AT LEAST 3 PROTEINS NOW PRODUCED BY RECOMBINANT DNA HAVE MEDICAL POSSIBILITIES...

HUMAN GROWTH HORMONE PREVENTS ONE TYPE OF DWARFISM. PEOPLE WHOSE GENETIC MAKE-UP WOULD OTHERWISE LEAVE THEM A BIT "SHORT," CAN GROW NORMALLY IF GIVEN ADEQUATE DOSES. SO FAR, DEMAND STILL EXCEEDS SUPPLY, BUT NOT FOR "LONG"!

I THINK I O.D.'D!

INSULIN, WHICH BREAKS DOWN SUGAR IN THE BLOOD, HAS LONG BEEN MADE BY OTHER MEANS... BUT SHOULD NOW BECOME MORE PLENTIFUL, AND POSSIBLY CHEAPER, MAKING LIFE EASIER FOR DIABETICS —

LET THEM EAT CAKE!

INTERFERON, THE VIRUS-FIGHTER, USED TO BE SO SCARCE IT COST A **TRILLION DOLLARS** AN OUNCE — BUT NOW IT'S MADE BY THE VATFUL BY TRILLIONS OF **E. COLI.** UNFORTUNATELY, NO ONE KNOWS EXACTLY WHAT TO DO WITH IT, THOUGH CLINICAL TRIALS CONTINUE AMID HIGH HOPES...

IT MAY CURE CANCER OR THE COMMON COLD!

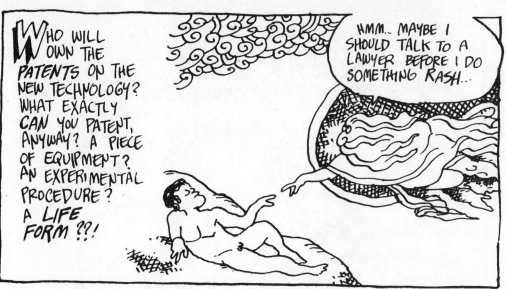

WHO WILL OWN THE **PATENTS** ON THE NEW TECHNOLOGY? WHAT EXACTLY CAN YOU PATENT, ANYWAY? A PIECE OF EQUIPMENT? AN EXPERIMENTAL PROCEDURE? A **LIFE FORM ??!**

HMM... MAYBE I SHOULD TALK TO A LAWYER BEFORE I DO SOMETHING RASH...

THIS QUESTION HAS ALREADY GONE TO THE SUPREME COURT, WHICH RULED THAT **NEWLY INVENTED LIFE FORMS** MAY BE PATENTED!

SOMEBODY OWES ME THREE BILLION YEARS' WORTH OF BACK ROYALTIES!!

SO THE PRESSURE IS ON... PROFESSORS ACCUSE EACH OTHER OF USING THE CAMPUS LABS FOR COMPANY BUSINESS... GRAD STUDENTS FIND THEIR PROJECTS CHANGED FOR NO APPARENT REASON... AND OF COURSE THERE'S THE JEALOUSY...

WHY SHOULD THOSE GENE SPLICERS GET RICH, WHILE I, DISCOVERER OF THE SNAIL DARTER'S MATING CYCLE, REMAIN POOR?

BUT FORGET ABOUT *MONEY*... WHAT ABOUT OUR *HEALTH* ?? FROM THE FIRST DAYS OF GENETIC ENGINEERING, PEOPLE HAVE WORRIED ABOUT BREEDING *MONSTERS* IN THE LAB !!

THE FEAR WAS THAT TAMPERING WITH E. COLI'S DNA MIGHT CREATE A SUPER-DEADLY GERM BY ACCIDENT.

REMEMBER, E. COLI LIVES IN THE HUMAN INTESTINE — IF A VIRULENT STRAIN SHOULD ESCAPE FROM THE LAB, THERE MIGHT BE NO STOPPING IT!! WHO'D HAVE THOUGHT FRANKENSTEIN'S MONSTER WOULD LOOK LIKE THIS?

GROWL

ACCORDINGLY, SCIENTISTS VOLUNTARILY ADOPTED GUIDELINES TO LIMIT POTENTIAL HAZARDS...

EMPLOYEES MUST WASH HANDS AFTER COMBINING GENES

SINCE THE EARLY DAYS, THE FEAR HAS FADED... THERE HAS BEEN NO SIGN OF A PROBLEM YET!

THE MOST ENCOURAGING THING IS THIS: THE STRAIN OF E. COLI USUALLY USED FOR CLONING GENES HAS GROWN SO "DOMESTICATED" DURING ITS YEARS IN THE LAB, THAT IT CAN NO LONGER SURVIVE IN THE HUMAN GUT!!

SO... MAYBE THERE'S NOTHING TO WORRY ABOUT... THOUGH IT'S TRUE THAT THE SAFEGUARDS ADOPTED BY UNIVERSITIES DON'T GENERALLY APPLY TO PRIVATE COMPANIES!!!

WHAT'S MUCH MORE LIKELY
IS THAT SOMEONE WILL MAKE A
DEADLY GERM *ON PURPOSE*.
WHO WOULD WANT TO DO
THAT, YOU ASK?

THE GENERALS HAVE BEEN KNOWN TO TURN NEW
TECHNOLOGY TO MILITARY USE, AND
THEY USUALLY FIND SCIENTISTS
TO OBLIGE...

WE CAN TAKE SOME COMFORT FROM THE FACT THAT BIOLOGICAL WARFARE IS BANNED BY INTERNATIONAL TREATY, BUT YOU NEVER KNOW...

LET ME TELL YOU ABOUT SOME BROKEN TREATIES!

IT'S A POLITICAL QUESTION RAISED BY A SCIENTIFIC ADVANCE — A FAMILIAR FACT OF 20TH CENTURY LIFE.

DOES THIS POTENTIAL FOR HARM MEAN THAT GENE SPLICING SHOULD BE STOPPED ?? ALMOST WITHOUT EXCEPTION, THE BIOLOGISTS SAY "NO." WHY REJECT THE MEDICAL ADVANCES ALONG WITH THE MILITARY USES ??

BESIDES, THE POISONS THAT COULD BE MADE THIS WAY ARE PROBABLY NO WORSE THAN THE ONES THAT ALREADY EXIST, WHILE MEDICAL ADVANCES PROMISE TO BE TRULY *REVOLUTIONARY*.

FORWARD!

YEAH... LET US REALIZE OUR HUMAN POTENTIAL...

ON THE VERGE

SO FAR, THE SUCCESSES IN THIS FIELD HAVE COME IN VIRUSES, BACTERIA, YEAST, AND PLANTS, BUT WE'RE GETTING MUCH CLOSER TO WORKING DIRECTLY WITH **HUMAN BEINGS.**

GAK! HUMANS? DISGUSTING!

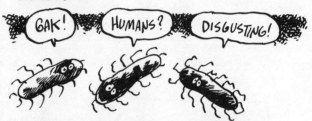

WHEN MAKING TESTS ON HUMANS, SCIENTISTS MUST APPLY A **DIFFERENT STANDARD** FROM THAT GOVERNING EXPERIMENTS ON ANIMALS OR BACTERIA.

NAMELY, IT'S SUPPOSED TO DO THE SUBJECT SOME GOOD!

THAT'S WHY WE KNOW SO WELL WHAT CAUSES CANCER IN *RATS...* HOW COULD YOU DO AN EXPERIMENT TO FIND THE CAUSES OF CANCER IN *HUMANS??*

ASK FOR VOLUNTEERS?

...WHICH IS TO SAY, EXPERIMENTS ON HUMANS STIR UP CONTROVERSY, A GOOD EXAMPLE BEING RECENT ATTEMPTS TO TREAT *THALASSEMIA.*

AS YOU RECALL, THIS CONDITION IS AN INABILITY TO MAKE *HEMOGLOBIN,* CAUSED BY A MISTAKEN "STOP" CODON IN THE MIDDLE OF THE GENE FOR ONE OF ITS CHAINS.

THALASSEMIA VICTIMS CAN SUFFER FROM ANEMIA, BONE DEFORMITIES, AND HEART PROBLEMS. THEY REQUIRE FREQUENT BLOOD TRANSFUSIONS TO SURVIVE, AND EVEN THEN THEY DON'T LIVE LONG.

WITH THE SUCCESS OF RECOMBINANT DNA, DOCTORS BEGAN TO HOPE THAT THE DISEASE COULD BE CURED BY SPLICING A *GOOD* GENE INTO A HUMAN CHROMOSOME.

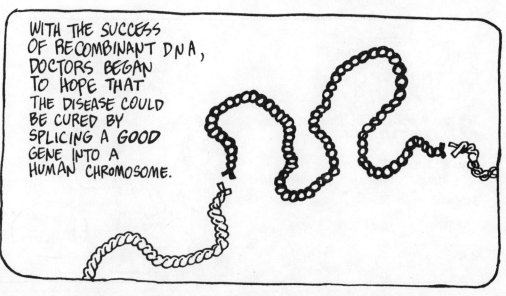

SOUNDS GOOD, EXCEPT THAT THE SAME APPROACH HAD ALREADY FAILED REPEATEDLY IN MICE. STILL, A TEAM OF DOCTORS FROM U.C.L.A. DECIDED TO TRY IT ON HUMANS ANYWAY..!

THEY REMOVED BONE MARROW CELLS FROM TWO PATIENTS' THIGH BONES. (REMEMBER, THESE DEVELOP INTO HEMOGLOBIN-PRODUCING RED BLOOD CELLS.)

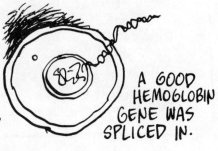

A GOOD HEMOGLOBIN GENE WAS SPLICED IN.

THE THIGH WAS IRRADIATED TO SLOW DOWN THE OLD MARROW (AND GIVE THE NEW CELLS THE EDGE).

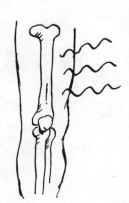

AND THE "ENGINEERED" CELLS WERE PUT BACK IN.

AND THE RESULT?

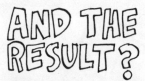 ABSOLUTELY NOTHING!

(SINCE THEN, THE EXPERIMENT HAS WORKED — IN MICE.)

≥SIGH≥ THERE GOES THE EXPERIMENT...

AND THE PATIENT...

THE DOCTORS TOOK A LOT OF FLAK FOR THIS EXPERIMENT.

SEVERAL OBJECTIONS WERE RAISED:

NOT EVEN A *PART* OF THE PROCEDURE HAD EVER WORKED IN ANIMALS. IT'S STILL NOT AT ALL CLEAR HOW TO INSERT A HUMAN HEMOGLOBIN GENE INTO A MAMMAL CELL IN SUCH A WAY THAT IT'S EXPRESSED IN ANY QUANTITY.

REGULATION IN MAMMALS IS STILL MURKY!

THE EXPERIMENT WAS *DISAPPROVED* BY U.C.L.A.'s COMMITTEE ON HUMAN SUBJECTS USE. HOWEVER, IT HAD BEEN APPROVED BY THE TWO HOSPITALS WHERE IT WAS CARRIED OUT (IN ITALY AND ISRAEL).

THE *RADIATION* CERTAINLY DIDN'T HELP THE PATIENTS. ON THE OTHER HAND, THEY BOTH FULLY UNDERSTOOD WHAT WAS BEING DONE, AND THEY GAVE THEIR CONSENT.

WERE THEY GRASPING AT STRAWS?

AFTERWARDS, THE DOCTORS WERE DISCIPLINED, ONE OF THEM LOSING HIS POSITION AS DEPARTMENT CHAIRMAN... SO YOU SEE—HUMAN EXPERIMENTS CAN BE DANGEROUS!

DANGEROUS TO DOCTORS, THAT IS!

NEVERTHELESS, THIS IS PROBABLY THE WAY THE FIRST GENETIC THERAPIES WILL BE DONE, BECAUSE BONE MARROW IS THE EASIEST TISSUE TO TRANSPLANT.

THERE ARE SEVERAL DISEASES THAT MIGHT BE TREATED THIS WAY:

THALASSEMIA, OF COURSE, ALTHOUGH THE UCLA EXPERIENCE SHOWS IT WON'T BE EASY.

IF AT FIRST YOU DON'T SUCCEED...

SICKLE-CELL ANEMIA IS A HEMOGLOBIN ABNORMALITY AFFECTING MAINLY BLACK PEOPLE. THIS WILL BE EVEN HARDER BECAUSE THE MUTANT GENE IS CODOMINANT, NOT RECESSIVE.

AND HEMOGLOBIN REGULATION IS STILL A PUZZLE...

HEMOPHILIA, DUE TO THE LACK OF A BLOOD PROTEIN, MIGHT BE THE EASIEST TO CURE!

YOU MAY BREATHE EASIER, MY PRINCES!

AND THERE ARE **IMMUNO-DEFICIENCY** DISEASES CAUSED BY RECESSIVE GENES IN BONE MARROW. AT PRESENT, PEOPLE WITH THESE DISEASES HAVE TO LIVE IN GERM-FREE ISOLATION CHAMBERS.

SO GET ON WITH THE RESEARCH!

OF COURSE, THERE ARE FEWER RESTRICTIONS ON PLANT AND ANIMAL EXPERIMENTS THAN ON HUMANS. (THIS BOTHERS SOME PEOPLE, BY THE WAY.)

SO PROGRESS HAS BEEN MORE RAPID AMONG PLANTS AND ANIMALS. ALREADY THERE ARE BREEDS OF COTTON, TOMATO, AND TOBACCO WITH AN ADDED BACTERIAL GENE THAT MAKES THEM POISONOUS TO INSECTS.

FORTUNATELY, TOBACCO WAS ALREADY POISONOUS TO HUMANS...

SCIENTISTS ARE EXCITED ABOUT TRANSGENIC ANIMALS — ANIMALS THAT CONTAIN A FEW GENES FROM ANOTHER SPECIES.

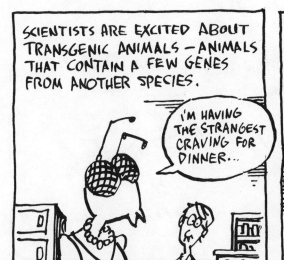

I'M HAVING THE STRANGEST CRAVING FOR DINNER...

ONE EXAMPLE ARE PIGS WITH BOVINE GROWTH HORMONE. THEY GROW FASTER AND LEANER, BUT ALSO HAVE OTHER PROBLEMS, LIKE ULCERS AND ARTHRITIS — SO YOU'LL HAVE TO WAIT FOR THAT "BORK" CHOP.

GET ME A PROTESTER!

TRANSGENIC PLANTS AND ANIMALS CAN PASS ON THEIR NEW GENES TO THEIR OFFSPRING, BECAUSE THE GENES ARE INSERTED AT A VERY EARLY STAGE OF DEVELOPMENT, ALLOWING THEM TO GET INTO SPERM AND EGG CELLS. PERFORMING THESE EXPERIMENTS ON HUMANS WOULD THEREFORE RAISE SOME HARD ETHICAL ISSUES.

YOU DON'T HAVE TO MAKE THE BABY PERFECT — JUST BETTER THAN ANYONE ELSE'S...

BUT WE'RE GETTING CLOSER. THERE ARE ALREADY LIVING "TEST TUBE BABIES" — FERTILIZED IN A TEST TUBE AND THEN, AFTER A FEW DIVISIONS, IMPLANTED IN THE MOTHER'S WOMB, WHERE THEY DEVELOPED NATURALLY.

HEY, MOM! HOW YA DOIN'?

WHAT WOULD THE MONK MENDEL HAVE TO SAY ABOUT **THIS?**

I'D SAY, "DON'T DROP THAT TEST TUBE!"

202

THE OBVIOUS NEXT STEP WOULD BE TO **ENGINEER** THE **EMBRYO** IN THE TEST TUBE...

THIS COULD RANGE FROM **GENE THERAPY** — FIXING SPECIFIC DEFECTS — TO... WHO KNOWS WHAT??

AT THE EXTREME, IT MAY BECOME POSSIBLE TO **CLONE PEOPLE**. THE EGG'S NUCLEUS WOULD BE REMOVED ALTOGETHER AND REPLACED WITH A NUCLEUS FROM ANOTHER PERSON.

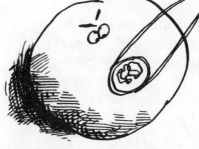

THIS EGG WOULD BE IMPLANTED IN A "MOTHER," TO WHOM IT WOULD BE GENETICALLY UNRELATED.

INSTEAD, THE LITTLE TYKE WOULD BE GENETICALLY IDENTICAL TO WHOEVER — OR WHATEVER — DONATED THE NUCLEUS.

SOUND FAR-FETCHED? WELL, SCIENTISTS HAVE ALREADY SUCCEEDED IN CLONING MICE AND FROGS...

THE TECHNIQUE MAKES IT POSSIBLE TO MAKE MULTIPLE COPIES OF LIVING INDIVIDUALS! IS THIS WHAT WE WANT TO BECOME, A WORLD OF CLONES??

WE·SEE·NOTHING· WRONG·WITH·IT!

YOU MIGHT WELL ASK: WHO WILL BE CLONED? WHO WILL DECIDE? WILL IT BE BASED PURELY ON MONEY? WILL IT BE LEGAL? WILL THERE BE PEOPLE-BREEDERS SELECTING THE MOST "FIT" FOR REPRODUCTION?

STAND ASIDE, WEAKLINGS!

THE LAST TIME ANYONE TRIED TO BREED A MASTER RACE, IT WAS AN UNHAPPY EXPERIENCE, TO SAY THE LEAST...

OR MAYBE WE'RE BEING TOO *GLOOMY*... MAYBE THE FUTURE WILL BE A GLORIOUS TIME WHEN PEOPLE WILL BE ENGINEERED TO FIT CLOTHES INSTEAD OF VICE VERSA !!

DO YOU HAVE 700 BLUE BLAZERS IN A 42 REGULAR?

WE CAN BREED PEOPLE WHO CAN WEAR SPIKE HEELS COMFORTABLY, OR WHO HAVE PINK HAIR !!

DEE-VOLVED !!

MAYBE WE CAN EVEN BE CLONED TO RESIST ECOLOGICAL DISASTER, LIKE THE DEPLETION OF ATMOSPHERIC OZONE !!

WE'LL INCREASE THE NUMBER OF GENES FOR SKIN PIGMENT TO SCREEN OUT COSMIC RAYS— YES, WE'LL INVENT BLACK PEOPLE !!!

DO TELL...

IT'S NOT ONLY OUR OWN GENES WE NEED TO WORRY ABOUT... THERE IS ALSO THE **GENETIC DIVERSITY** OF THE ENTIRE PLANET... (IT LOOKS SOMETHING LIKE A GIANT CELL, DOESN'T IT?)

IT'S HARDLY NEWS THAT ALL LIFE IS INTERDEPENDENT... GORILLA EATS BANANA; BANANA EATS CHEMICALS FROM THE SOIL; SOME OF THE CHEMICALS GET THERE FROM BACTERIAL ACTION; OTHER BACTERIA AID THE APE'S DIGESTION; STILL OTHERS BREAK DOWN ITS WASTE PRODUCTS, ETC ETC ETC...

BUT WE HUMANS

WITH OUR EXPLODING POPULATION, RESOURCE-HOGGING, MODERN AGRICULTURE, AND POLLUTION, ARE CHANGING THE ENVIRONMENT SO DRASTICALLY THAT HUNDREDS OF PLANT AND ANIMAL SPECIES GO EXTINCT EVERY YEAR.

206

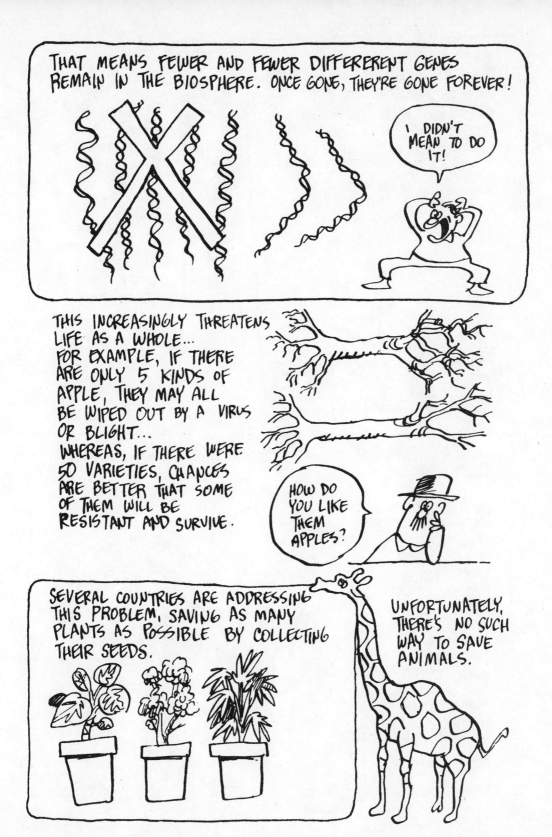

THAT MEANS FEWER AND FEWER DIFFERERENT GENES REMAIN IN THE BIOSPHERE. ONCE GONE, THEY'RE GONE FOREVER!

I DIDN'T MEAN TO DO IT!

THIS INCREASINGLY THREATENS LIFE AS A WHOLE... FOR EXAMPLE, IF THERE ARE ONLY 5 KINDS OF APPLE, THEY MAY ALL BE WIPED OUT BY A VIRUS OR BLIGHT... WHEREAS, IF THERE WERE 50 VARIETIES, CHANCES ARE BETTER THAT SOME OF THEM WILL BE RESISTANT AND SURVIVE.

HOW DO YOU LIKE THEM APPLES?

SEVERAL COUNTRIES ARE ADDRESSING THIS PROBLEM, SAVING AS MANY PLANTS AS POSSIBLE BY COLLECTING THEIR SEEDS.

UNFORTUNATELY, THERE'S NO SUCH WAY TO SAVE ANIMALS.

PERHAPS GENETIC ENGINEERING WILL BE ABLE TO HELP BY CREATING NEW COMBINATIONS, BUT THIS IS STILL IN THE FUTURE...

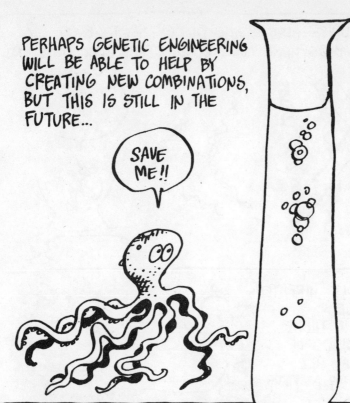

ON THE OTHER HAND, THE POSSIBILITIES FOR GENETIC ENGINEERING WILL BE LIMITED BY THE LIMITED NUMBER OF ALLELES LEFT TO RECOMBINE.

WE FIND OURSELVES CONFRONTED BY OUR OWN AWESOME POWERS.

ON THE ONE HAND, WE FACE THE BLIND POWER THAT STRIPS FORESTS, ERODES THE SOIL, TURNS MARGINAL FARMLAND INTO DESERT AND DEPLETES THE HEALTHY DIVERSITY OF THE GENE POOL...

ON THE OTHER HAND, WE MUST DEAL WITH THE GROWING POWER OF GENETIC ENGINEERING. IT PROMISES — OR THREATENS — TO ALTER THE VERY NATURE OF HUMANITY. IT RAISES QUESTIONS WHICH WE BARELY HAVE A VOCABULARY TO DISCUSS, MUCH LESS SOCIAL AND POLITICAL INSTITUTIONS TO DECIDE.

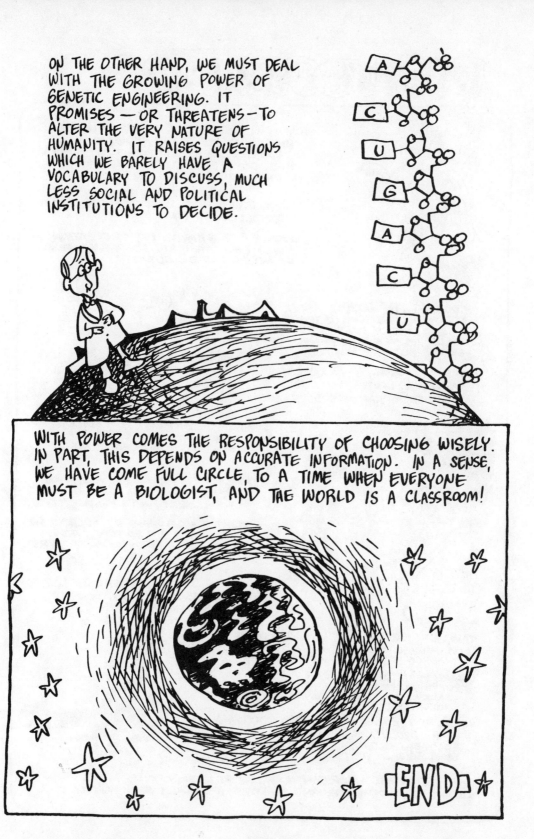

WITH POWER COMES THE RESPONSIBILITY OF CHOOSING WISELY. IN PART, THIS DEPENDS ON ACCURATE INFORMATION. IN A SENSE, WE HAVE COME FULL CIRCLE, TO A TIME WHEN EVERYONE MUST BE A BIOLOGIST, AND THE WORLD IS A CLASSROOM!

END

BIBLIOGRAPHY

STUBBE, H., *HISTORY OF GENETICS FROM PRE-HISTORIC TIMES TO THE REDISCOVERY OF MENDEL'S LAWS*, M.I.T. PRESS, 1972. HARD TO FIND, BUT A FINE SCHOLARLY HISTORY OF GENETICS TO 1900.

DUNN, L.C., *A SHORT HISTORY OF GENETICS*, McGRAW-HILL, 1965. MORE PRE-1939 GENETICS. GOOD PIX.

JUDSON, H.F., *THE EIGHTH DAY OF CREATION*, SIMON & SCHUSTER, 1979. READABLE HISTORY OF MOLECULAR BIOLOGY.

WATSON, J.D., *THE DOUBLE HELIX*, ATHANEUM, 1968. ONE OF THE DISCOVERERS OF DNA'S STRUCTURE TELLS HIS STORY. FLIPPANT AND SEXIST, BUT FASCINATING.

SAYRE, A., *ROSALIND FRANKLIN AND DNA*, NORTON, 1978. AN ANTIDOTE TO WATSON'S BIAS.

CURTIS, H., *BIOLOGY, 2ND EDITION*, WORTH, 1975. A GOOD GENERAL BIO TEXT, FOR MORE ON MOLECULES AND CELLS.

AYALA, F.J., + KEIGER, J.A., *MODERN GENETICS*, BENJAMIN CUMMINGS, 1980. ONE OF MANY UP-TO-DATE TEXTS.

STENT, G. + CALENDAR, R. *MOLECULAR GENETICS, 2ND EDITION*, FREEMAN, 1978. ALL THE DETAILS. (THE FIRST EDITION, BY STENT ALONE, IS A CLASSIC, THOUGH DATED.)

WATSON, J.D., *MOLECULAR BIOLOGY OF THE GENE, 3RD EDITION*, W.A. BENJAMIN, 1976. MORE DETAILS.

CAVALIERI, L.F., *THE DOUBLE-EDGED HELIX*, COLUMBIA U. PRESS 1981; SUBTITLED "SCIENCE IN THE REAL WORLD".

CHARGAFF, E. *HERACLITEAN FIRE*, ROCKEFELLER U. PRESS, 1978. A CRANKY MEMOIR, BUT MAYBE WE SHOULD LISTEN TO HIM!

WADE, N., *THE ULTIMATE EXPERIMENT: MAN-MADE EVOLUTION*, WALKER & CO, 1977. RECOMBINANT DNA, BY ONE OF OUR BEST SCIENCE WRITERS.

ALSO: *SCIENTIFIC AMERICAN* MAGAZINE REGULARLY PRINTS ARTICLES ON RECENT DEVELOPMENTS, AND SO DOES YOUR DAILY NEWSPAPER!

INDEX

Adenine, 120, 122
AIDS virus, 156
Alleles, 42–50, 54
    co-dominant, 162
    combinations of, 54
    recessive, 81
Amino acids, 108–111
Anthers, 31
Antibodies, 177–178
Anticodon, 136–137
Aristotle, 14–15
Asparagine, 109
Assortment, independent, principle of, 48, 70
Attenuation, 172–174
Avery, Oswald, 118–119

Bacteria, 25
Baldness, hereditary, 91–93
Base pairs, 122–123
    sequence of, 130
Bases, 106, 107
Beadle, George, 114
Bibliography, 210
Biological warfare, 194–195
Blood groups, 162
Body cells, 83
Bone marrow cell, 163
Bonellia marine worm, 90
Bovine growth hormone, 201
Breeding, selective, 6

Camerarius, 30
Cancer, 83
Carbon, 104
Cell division, 57
Cells, 56–64, 97–101
    turning into other kinds of cells, 163
    types of, 98–99, 164
Cellulose, 105
Centromere, 59
"Chaperone" protein, 129
Chargaff, Erwin, 121

Chromosome mapping, 69–78
Chromosome number, 60
Chromosomes, 58–70, 102
    genes on, 67, 69–71
    "recombinant," 76
    in sperm and egg, 61–62
    X and Y, 85–89, 91–93
"Clipping" enzyme, 127
Cloning genes, 185–188, 193
Cloning people, 203
Co-dominant alleles, 162
Codons, 133–137
    complementary, 136–137
Color-blindness, 91–92
Complement, 124
Complementarity, principle of, 124, 128
Complementary codon, 136–137
Correns, Carl, 65–66
Crick, Francis, 122–125
Crops, productive, 7
Crossing over, gene, 71, 76–77
Crossing square, 44–45, 48, 73–74, 93
Cysteine, 109
Cytosine, 120, 122

Darwin, Charles, 55
Deoxyribonucleic acid, *see* DNA *entries*
Deoxyribose, 106
DeVries, Hugo, 65–66
Digestive enzymes, 112
Diploid organisms, 68, 89
Diversity, genetic, 206–208
Division, cell, 57
DNA (deoxyribonucleic acid), 107, 119, 181
    "junk," 147
    recombinant, 180, 184, 188
    repetitive, 152–153
    "selfish," 153
    sequence of, 129–130
DNA replication, 125–128
Dominant trait, 40–48
    examples of, 53
Double helix, 123–125

E. coli, see *Escherichia coli*
Eco-R1 enzyme, 182–183
Egg, 29
  chromosomes in, 61–62
  mammalian, 27–29
Empedocles, 16
Endonuclease, restriction, 182
Engineering, genetic, 179–195, 206–209
Environment, 206, 208
Enzyme Eco-R1, 182–183
Enzymes, 112–116, 129, 181
  "clipping," 127
  digestive, 112
  genes and, 114–116
  "inducible," 171
  "snipping," 126–127
Escherichia coli (E. coli), 100–102,
    185–186, 192–193
  histidine and, 172–174
  lactose and, 166–171
  replication, 126
Eucaryotes, 144–148
Experiments on humans, 196–199,
    202–205
Extinction, species, 206–208

Fertility magic, 9
Fertilization, 29
  plant, 31
Flippase gene, 176
Flowers, 30–31
Franklin, Rosalind, 121

Gametes, 61
Gene cloning, 185–188, 193
Gene expression, 174
Gene mapmaking, 69–78
Gene mutation, 79–83, 158–161
Gene regulation, 164–178
Gene splicing, 180–184
Gene suppression, 174
Gene swapping, 71, 76–77
Gene therapy, 203
Generation, spontaneous, 20–23
Genes, 42, 54, 96
  bacterial, 201
  on chromosomes, 67, 69–71
  dominant and recessive, 40–48, 53
  enzymes and, 114–116
  flippase, 176
  jumping, 175–178

  partial, 178
  sex-linked, 91–95
Genetic code, 134–137, 145
Genetic code table, 135
Genetic diversity, 206–208
Genetic engineering, 179–195, 206–209
Genetic engineering company, 189
Genetic research, peas in, 38–49
Geneticists, 32
Genetics, practical, 6
Genotype, 49
Gilbert, Walter, 171
Glucose, 104
Glycine, 109
Gonick, Larry, 215
Griffith, Fred, 116–117
Growth hormone, human, 188
Guanine, 120, 122
Guanine "cap," 146

Haploid organisms, 68, 89
Harvey, William, 27–28
Helix, double, 123–125
Hemoglobin, 108, 158–159, 197–199
Hemophilia, 91, 94–95, 200
Hereditary baldness, 91–93
Hereditary traits, 54
Heredity, theories of, 12
Hertwig, Oscar, 29
Heterozygote, 49–52
Hippocrates, 13
Histidine, 172–174
Homolog, 64
Homologous pairs, 62–64, 67
Homozygote, 49, 51
Hooke, Robert, 56
Human growth hormone, 188
Humans, experiments on, 196–199,
    202–205
Hybrids, 33–34
  Mendel and, 39
Hydrogen, 104
Hydrogen bonding, 122

Immunodeficiency diseases, 200
Independent assortment, principle of, 48,
    70
"Inducible" enzymes, 171
Inheritance, 11
Insulin, 188
Interferon, 188
Introns, 148

Jacob, François, 166–167, 169
Jacob's flock, 9–10, 33, 50
Jumping genes, 175–178
"Junk" DNA, 147

Karyotype, 88
Kleinfelter's female, 87

Lactose, 166–171
Leader sequence, 173
Leewenhoek, Anton van, 23–26
Leucine, 109
Ligase, 184
"Like begets like" notion, 6, 10
Lipids, 105
Loops, mRNA, 147–148

Macromolecules, 104–105
Magic, 8–10
    fertility, 9
    sympathetic, 10
Mammalian egg, 27–29
Mapmaking, gene, 69–78
Marine worm *Bonellia*, 90
Marrow cell, 163
Mating, 5
Meiosis, 63–64, 86
    faulty, 87
Mendel, Gregor, 37–48, 55
    hybrids and, 39
    principal results, 54
"Messenger molecule," 131
Messenger RNA, *see* mRNA *entries*
Methionine, 140
Mice, "waltzing," 11
Microscope, 23–24
Mitosis, 59–60
Mold, *Neurospora*, 114–115
Molecules, 104–105
Monod, Jacques, 166–167, 169
mRNA (messenger RNA), 133, 139, 142, 143, 149, 173
    editing of, 157
mRNA loops, 147–148
Müller, Hermann, 82
Müller-Hill, B., 171
Mutagens, 82
Mutation, gene, 79–83, 158–161

Nerve cells, 98
Neurospora mold, 114–115
Nirenberg, Marshall, 134

Nitrogen, 104
Nuclear membrane, 146
Nucleic acids, 106–107, 154
Nucleosome core, 150
Nucleotides, 106, 126–127, 148
Nucleus, cell, 58, 144

Operator, 169–170
Operon, 167, 173–174
Origin, 126
Ovary, 31
Oxygen, 104

Partial genes, 178
Pasteur, Louis, 57
Patents, 191
Peas in genetic research, 38–49
People
    cloning, 203
    primitive, 1–4
Peptide, 110
Perutz, Max, 108
Phenotype, 49
Phenylalanine, 109, 134
Phosphate, 104, 106, 107
Phosphorus, 104
Plant sex, 30–31
Plasmids, 185
Pneumococcus, 116–117
Pollen, 31
Pollination, 8
Poly-A tail, 146
Polypeptide, 110
Polyploid organisms, 68
Polysaccharides, 105
Primitive people, 1–4
Principle
    of complementarity, 124, 128
    of independent assortment, 48, 70
Procaryotes, 144–145
Productive crops, 7
Promoter region, 167–169
Protein chain, 110
Protein coat, 154
Protein synthesis, 133
Proteins, 106, 108–113, 129
    "chaperone," 129
Protozoa, 57, 90

Recessive allele, 81
Recessive genes, 200
Recessive traits, 40–41, 161
    examples of, 53

"Recombinant" chromosomes, 76
Recombinant DNA, 180, 184, 188
Red blood cells, 98, 163
Redi, Francesco, 22
Repetitive DNA, 152–153
Replication
  DNA, 125–128
  E. coli, 126
Repressive tolerance, 157
Repressor, 169–171
Reproduction, *v,* 3–5
Restriction endonuclease, 182
Retrovirus, 156
Ribonucleic acid, *see* RNA *entries*
Ribose, 106, 132
Ribosomal RNA (rRNA), 138, 143
Ribosome, 138–143
RNA (ribonucleic acid), 107, 132
  messenger, *see* mRNA *entries*
  ribosomal, 138, 143
  transfer, *see* tRNA
RNA polymerase, 133, 167–168, 173–174
rRNA (ribosomal RNA), 138, 143

Selective breeding, 6
"Selfish" DNA, 153
Semen, 13, 14, 25–26
Sequence
  of base pairs, 130
  of DNA, 129–130
Sex, 4–5
  determination of, 84–87
  plant, 30–31
Sex-linked genes, 91–95
Sickle-cell anemia, 200
"Snipping" enzyme, 126–127
Socrates, 12
Somatic cells, 83
Species extinction, 206–208
Sperm, 25–26, 28, 29
  chromosomes in, 61–62
  spindle, 59, 63
Splicing, gene, 180–184
Spliceosome, 147
Spontaneous generation, 20–23
Sports, 34–35, 52
Square, crossing, 44–45, 48, 73–74, 93
Stable varieties, 33
Starch, 105
Stigma, 31

"Stop" signals, 135, 141
Sugars, 104–107
Sulfur, 104
Sunbathing, 83
Sutton, William, 62, 66
Sympathetic magic, 10

Tatum, Edward, 114
"Test tube babies," 202–203
Tetrads, 63
Thalassemia, 159, 161, 197, 200–201
Thymine, 120, 122
Transcription, 133, 148, 173
Transfer RNA, *see* tRNA
Transforming factor, 118
Transgenic animals, 201, 202
"Translator" molecules, 131
Transposons, 177
tRNA (transfer RNA), 136–137, 138,
  139–141, 143
Tryptophan, 109
Tschermak, Erich von, 65–66
Turner's female, 87

Ultraviolet light, 83
Uracil, 132

Varieties, stable, 33
Vector, 185
Victoria, Queen of England, 94–95
Viruses, 154–157

"Waltzing" mice, 11
Warfare, biological, 194–195
Water, 104
Watson, James, 122–125
Wheelis, Mark, 215
Worm, marine, *Bonellia,* 90

X chromosome, 85–89, 91–93
X-rays, 82
Xenophon, 11
XXY syndrome, 87
XYY karyotype, 88

Y chromosome, 85–89, 91–93
Yeast, 187

Zygote, 61

ABOUT THE AUTHORS:

LARRY GONICK IS THE AUTHOR OR CO-AUTHOR OF MANY BOOKS OF GRAPHIC NON-FICTION ON SCIENTIFIC AND HISTORICAL SUBJECTS. A GRADUATE OF HARVARD IN MATH, HE DROPPED OUT OF GRADUATE SCHOOL TO PURSUE SOMETHING REALLY DIFFICULT: RENDERING INFORMATION IN LITTLE PICTURES. HE LIVES WITH HIS FAMILY IN SAN FRANCISCO.

MARK WHEELIS, WHEN NOT CLIMBING ROCKS OR RAFTING RIVERS, IS SENIOR LECTURER IN MICROBIOLOGY AT THE UNIVERSITY OF CALIFORNIA AT DAVIS. BESIDES TEACHING NUMEROUS BIOLOGY COURSES, HE HAS WRITTEN MANY RESEARCH PAPERS AND IS CO-AUTHOR OF THE STANDARD TEXTBOOK *THE MICROBIAL WORLD.* HE LIVES IN DAVIS WITH HIS WIFE, CHILDREN, DOG, AND MICROBES.